# ADVENTURESOME SOUL

*Life Stories of an Adrenaline Junky*

SPENCER NICHOLL

WestBow Press
A DIVISION OF THOMAS NELSON
& ZONDERVAN

Copyright © 2022 Spencer Nicholl.

All rights reserved. No part of this book may be used or reproduced by any means, graphic, electronic, or mechanical, including photocopying, recording, taping or by any information storage retrieval system without the written permission of the author except in the case of brief quotations embodied in critical articles and reviews.

This book is a work of non-fiction. Unless otherwise noted, the author and the publisher make no explicit guarantees as to the accuracy of the information contained in this book and in some cases, names of people and places have been altered to protect their privacy.

WestBow Press books may be ordered through booksellers or by contacting:

WestBow Press
A Division of Thomas Nelson & Zondervan
1663 Liberty Drive
Bloomington, IN 47403
www.westbowpress.com
844-714-3454

Because of the dynamic nature of the Internet, any web addresses or links contained in this book may have changed since publication and may no longer be valid. The views expressed in this work are solely those of the author and do not necessarily reflect the views of the publisher, and the publisher hereby disclaims any responsibility for them.

Any people depicted in stock imagery provided by Getty Images are models, and such images are being used for illustrative purposes only.
Certain stock imagery © Getty Images.

Unless otherwise indicated, all Scripture quotations are taken from the Holy Bible, New Living Translation, copyright © 1996, 2004, 2015 by Tyndale House Foundation. Used by permission of Tyndale House Publishers, Carol Stream, Illinois 60188. All rights reserved.

Scripture quotations marked (NIV) taken from The Holy Bible, New International Version® NIV® Copyright © 1973 1978 1984 2011 by Biblica, Inc. TM Used by permission. All rights reserved worldwide.

ISBN: 978-1-6642-5538-8 (sc)
ISBN: 978-1-6642-5539-5 (hc)
ISBN: 978-1-6642-5537-1 (e)

Library of Congress Control Number: 2022900544

Print information available on the last page.

WestBow Press rev. date: 01/04/2023

"If I could choose one man for my sons to emulate, it would be Spencer Nicholl. He is a one-of-a-kind blend of deeply spiritual leader, missionary explorer, common sense philosopher, and rugged outdoorsman. In *Adventuresome Soul*, Spencer takes us along on his wild journey from reluctant enthusiast to true believer. He lives out what the wannabes can only dream of. I was deeply touched, thoroughly entertained, as well as moved to action by his writing. *Adventuresome Soul* merits shelf space in any outdoorsman's library. On second thought, DO NOT relegate this book to the shelf; that *would* be a sin! Once finished reading this book, place it in the hands of an adventurous non-believer. Then you too will be emulating Spencer—by leading others to Christ."
—Randy Ulmer
Member of Archery Hall of Fame, Bowhunter's Hall of Fame and renowned big-game hunter

"*Warning*: Reading this book could be hazardous to mundane living. If you're looking for a shot of 'adventure espresso,' Spencer will quickly become your favorite barista as he serves up story after story of raw adventure guaranteed to reawaken you to something God has hardwired into each one of our souls."
—Doug Pollock
Speaker and author of *God Space* and coauthor of
*Irresistible Evangelism*

From the mountains of Colorado to the ice of Siberia, Spencer Nicholl and I have shared more adventures than I can count. *Adventuresome Soul* is one man's soul journey from a worship of creation to a life saturated by the Creator's heart and love for people *and* the beautiful artwork of His creation. We have hunted bears, moose, cougars, mountain goats, deer, elk and pronghorn together! Spencer's skills in the outdoors sets him apart from most but his greatest skill is taking the pure milk of God's wisdom and making it accessible to all. He is a pastor's pastor and to call him friend is one of my greatest gifts in life. The true stories you are about to read will read you.

—John Mark Lamb

Professor at Colorado Christian University and CRU staff member for 45 years

Men like Ernest Shackleton and Teddy Roosevelt had the privilege of living in the golden age of exploration, uncovering geological secrets, and making first boot tracks on the South Pole and first voyage of the River of Doubt. In many ways, we are living in the final days of another golden age and have the honor of bringing the message of God and his Kingdom to far-flung communities who have never been offered Christ's love in this way before. Spencer Nicholl is one of the rare men who for decades has given a sacrificial and heroic *yes* to traveling the globe as an ambassador of the fullness of Life to those in need. Gathering antics from Mongolia to Iraq, Sudan to Russia, and many geographies in between, Spencer has shared with us a one-of-a-kind memoir filled with God-fueled stories that will beckon even the timid in heart to take one more step into the kind of adventures that make the souls of men come alive. With humor, wit, and moments of faith that could move mountains, **Adventuresome Soul** invites us behind the curtain into Spencer's global Kingdom adventures, leaving us to scratch our heads in wonder at the prevailing wildness and goodness of the heart of God. It's been said that when a man walks with God from the heart, every day is an adventure. Spencer breaks the limits of what's possible with God and invites us to do the same.

Morgan Snyder

Author, *Becoming a King*

Vice President, Wild at Heart

Founder, BecomeGoodSoil.com

# CONTENTS

Acknowledgments ..................................................................... ix
Preface ........................................................................................ xi
Introduction ............................................................................ xiii

| | | |
|---|---|---|
| Chapter 1 | Adventure is Caught ........................................... 1 |
| Chapter 2 | North to Alaska .................................................. 7 |
| Chapter 3 | Bush or Bust ...................................................... 13 |
| Chapter 4 | Father Spencer? ................................................. 25 |
| Chapter 5 | The Return of Two Prodigals ........................... 31 |
| Chapter 6 | Guided by God .................................................. 37 |
| Chapter 7 | Goodbye and Sam Bino! ................................... 44 |
| Chapter 8 | The Price is Right ............................................. 49 |
| Chapter 9 | Want to lose some weight? ............................... 56 |
| Chapter 10 | The world is round and has IVs ...................... 67 |
| Chapter 11 | My Second ......................................................... 76 |
| Chapter 12 | Siberia and a baby ............................................ 81 |
| Chapter 13 | Always Another Way ........................................ 88 |
| Chapter 14 | Joys and Pain .................................................... 97 |
| Chapter 15 | Planes, Trains, and Taxis ................................ 108 |
| Chapter 16 | The Wild West ................................................. 116 |
| Chapter 17 | Door or Barf Bag ............................................. 122 |
| Chapter 18 | Missionary meets 007 ..................................... 127 |
| Chapter 19 | Fear can lead to faith ...................................... 134 |
| Chapter 20 | Don't Run. Hide! ............................................. 149 |
| Chapter 21 | Wanna Go to Iraq? .......................................... 159 |
| Chapter 22 | The Adventurer ............................................... 167 |

Conclusion – The Adventure Continues ............................. 171

# Acknowledgments

To Annette, Andrew, and Kate:

You three inspire and encourage me. Where would I be without you? And to those—you know who you are—who have been such a part of my journey, both the adventurous souls and the faith-filled souls, thank you! Some of you have led me; others have infused me with the courage I lacked. Some of you have locked arms with me and been partners along the way. None of us walk alone, and the race we are running was never meant to be an individual event. Thank you, team!

# Preface

This book was written in part because I could no longer ignore the advice and encouragement from a handful of people to write it. They gave advice and encouragement to record the stories, to tell them in a way so they would last, and to tell them for my kids and their kids. This book is also written to reveal the power of God to draw a person to himself and extend an invitation into a faith journey packed with adventure. This journey, guided by the Father, has led me around the globe to places and people I had only dreamed of—from Alaska to Mongolia and Siberia, from Sudan to Iraq, to name a few. God has coaxed me to keep coming closer to him, and each step has revealed a grander and more purposeful adventure. Next to my computer, where I sit writing, is a collage of photos taken with Annette, my partner in this adventure and the love of my life. They include the two of us at Petra and Machu Pichu; in a boat on the canals of Brugge, Belgium; at a waterfall in Cameroon; and along the banks of the Jordan River.

As you read this, hopefully you will see not only adventure but also the power of God to change a person. Each step I have taken closer to him is a journey of going deeper into a relationship with the God of the universe and deeper into a life lived for his purpose. I sincerely hope that through these pages you will laugh, maybe learn, and listen more closely for God's voice. My prayer, as you read, is that your desire for the author and the perfecter of your faith will increase and inspire a willingness to step closer and into his adventure for you.

# Introduction

It was only twelve miles on US Highway 50 from the tiny town of Parlin to Gunnison. But Annette, my wife, covered it in record time. I tried not to look over at the speedometer, since I didn't really want to know how fast she was driving, and I didn't want to distract her, which is also why I just turned and looked out the window. Tears puddled in the corners of my eyes and started to run down my cheeks, and I didn't want those to be a distraction either. I knew they would if she saw them or my red face.

If you are driving upwards of ninety miles per hour, you really need to focus singularly on that task. My wife was doing a great job. I, on the other hand, wasn't doing so well. We had left our house about fifteen minutes earlier with me behind the wheel. We were on our way to meet our daughter, Kate, for a bite of dinner before attending the "Night of Lights" on Main Street in Gunnison. Turning on all the Christmas lights that adorned the light poles for several blocks and lighting the huge Christmas tree were an annual event. Lots of people packed the streets, and it really was a great time to be with our community. It was a street party of sorts, with folks so bundled up you had a hard time recognizing anyone.

The first time I felt the pain in my chest was several months before this. In fact, it was back in September, the day I was packing out the elk I had just taken with my bow, a strenuous activity indeed. But it had been made easier by the hype of the hunt, leftover adrenaline, and the natural high that came from being in the Colorado mountains with good friends. The pain in my chest must have been the result of a pulled pectoral muscle, I reckoned. *What else could it be?* I wondered since I was in good shape for fifty-two years old, and I was acclimated to the high altitude. As soon as my heart rate went back to normal, the pain was gone.

Over the next couple of months, I noticed that same pain off and on, usually when I was exerting myself by hiking or exercising. In fact, just the day before this, I had felt the pain again and thought, *Wow, that pec muscle just won't heal.* Then another thought hit me. *What if this is something else? Lord, this wouldn't be a good place to have a heart attack, since I'm a mile from the road and hiking with my good friend.* Right after my thought/prayer, I turned back downhill and slowly walked back to the pickup. The slight descent allowed my heart rate to return to normal, and once again, I felt fine.

If we were going out to dinner before the big celebration, I knew I'd better get some exercise to justify the extra calories, I reasoned. So I did my usual workout of push-ups, hanging sit-ups, and some weightlifting, the same workout I had done for about a year. If my heart rate climbed to the 150-beats-per-minute range for a half hour, then I deemed it a sufficient workout. I felt rotten after the workout. A hot shower would make me feel better, I thought. But it didn't.

I dressed and went to start the car to let it warm up, a practice you understand if you live in a cold climate like Gunnison. This allows the engine to warm before driving; it is also a chance to bring the interior temp up to something comfortable. After about ten minutes, I told Annette I would be in the car, waiting for her. She jumped in, and down the valley we drove. I couldn't get the chest pain to subside, even though my workout had been over for thirty minutes and my heart rate had returned to normal twenty-nine minutes ago. Both arms felt like they weighed an extra fifty pounds, and the pain seemed more widespread now. In fact, I felt as if three fat guys had parked their butts on my chest.

God often has a way of directing us and our steps, and we seldom recognize this at the time. But then hindsight brings into focus those circumstances, and we see his hand guiding us. This would be one of those times as we looked back at the event a couple of days later. At the prompting of Annette, I pulled over to let her drive, the first good, God-directed decision. She asked whether we should just forget the evening and return home. I said, "No, I'm sure this will pass, and I'll feel better by the time we get to town," the second good, God-directed decision.

Annette grabbed my hand, and we prayed as she drove down the county road toward the highway. There were only a couple of miles, but during those few minutes, I felt noticeably worse. So at about the time we turned on to Highway 50, she looked at me and said, "We need to go to the hospital, don't we?"

I mustered, "Yes, and go fast!"

My mind seemed clear. My thoughts were concise and focused, though you might call my response time after two and a half months of warning signs rather slow, and I would agree, especially for someone who had completed EMT training and served on our local volunteer fire department. Guilty as charged! But now things were clear. I was fully aware that I was having a heart attack. And despite the discomfort, the clarity of my thoughts and prayers was crystal. I didn't find myself praying for God to spare my life. I knew he had my days numbered, and if this was my day, then I was ready. I did find myself praying for my family, asking God to give them the grace they would need, to draw them close to him and each other in my absence. That turned on the tear faucet. But Annette didn't need to see me pouring out my heart and tears to God right now. She needed to drive like Mario Andretti!

What is it about near-death experiences that bring clarity? What happens to the fog during those moments? There seemed to be an element of truth to the saying "My life flashed before my eyes." I didn't see it flashing or even speeding by, but I suddenly had a flood of memories from early in life right up until they collided with my present. And that flood wasn't sad or scary or even regretful. The rush of different memories brought a smile and deep gratitude for the incredible life God had allowed me to live. I think it was William

Wallace who reportedly said, "Every man dies, but not every man really lives." I felt such a connection to that statement right then. The flood now turned on the emotions. I felt like God had allowed me to *really* live. He had brought me from death to life and had breathed his life into me. He'd allowed me to live out an incredible journey he had marked out for me, a journey that, as it turned out, wasn't over yet. The journey must have a few chapters left. The car ride that day didn't last long, thanks to Annette, but it allowed me enough time to pour myself out to God and reminisce about the years of adventure now coming at me in speedy, chronological order.

Modern medicine is incredible. I'm so thankful for good drugs, good tests, and the ability to run a stent up my femoral artery all the way into the clogged portion of the artery in my heart. It is amazing how much better things flow when you put a "culvert in the ditch." I felt so much better and was back home in about forty hours, thankful for great health care. This too was an adventure—not one of my choosing but still an adventure. I'm convinced that the greatest adventure in life is one that connects our hearts to the heart of God. (There is really no need for travel, an adrenaline rush, or epic endeavors as requirements for adventure.) All a person needs is to couple his or her heart to God's and hang on. In this earthly-divine relationship, there is plenty of risk, reward, and the unknown. But while growing up, forty years before this heart attack, I was unaware of the source of life or adventure, so I was seeking the epic and the rush the only way I knew how.

Chapter 1

# Adventure is Caught

There is part of each of us, I believe, that longs for adventure. It's something we are born with that begins to show up in the early years of life. So many of the books and movies we love are wrought with adventure. And if you grew up in a family like mine, adventure was a normal part of many conversations. Sitting around the dinner table while growing up was like having a front-row seat at some storytelling performance. It was usually my dad who was on stage.

Adventure and the ability to tell the story about it held an unspoken value in our family. I knew at an early age that I must be destined for great adventures. Even my name, Spencer, had significance to that end. I was named after a sporting goods store in Denver, Colorado. It was apparently my dad's favorite place to shop and spend his paychecks as a single man. I wasn't named after the shopping part of the experience, thank God. I was named more for what the store sold, the possibility of outdoor adventure.

Surely this meant that I, too, would have grand stories to pass on to the generations after me. But for now, I was content to listen to others' adventures and fantasize about what mine might be. Because of our love for the outdoors through hunting, fishing, and exploring, these yet-to-be-realized adventures would most certainly include these very activities. My time would come. But for now, I knew adventure was just a part of me and of my family. It just naturally seemed to seep into my being. It rubbed off a little more each time my father regaled us with a story from his youth or one from a recent hunt he had been on. I think you could catch that spirit for adventure easier than you could catch a cold or flu around our dinner table.

I was six years old when our family moved to Ohio City, Colorado. My brother, Matt, was seven; and and sister, Heidi, was three. We moved four hours southwest from Clear Creek County, just west of Denver. My dad worked as a lineman for Public Service Company, and my parents also ran Clear Creek Guest Ranch for my grandfather. The ranch consisted of a handful of rustic rental cabins a couple of miles downstream from Georgetown. We were tucked into the Rocky Mountains during a quieter time of the late '60s.

My grandfather, who lived near Denver, also owned several mines in the area. During the latter part of the 1800s and early 1900s, this area was congested with miners. Most of the mines had long since been closed. But my grandfather still worked one of his claims during my dad's youth. Most weekends found my dad, uncles, and grandpa working a hard-rock silver mine. Hard work for sure, but with it came the hope of great reward, which seem to be two elements commonly found in most good adventures.

There has been a westward movement of hardy folks over the last few centuries in this country. Moving west was somewhat synonymous with the unknown and taking a risk, two more elements that must be part of an adventure. And while my parents' move of only four hours west by car wasn't quite the same as a covered wagon on the Oregon Trail, it did have some unknowns and risks. There was, at least to my father, the need to see and learn new country that a move to the western slope would bring. Interstate 70 was being built a couple of hundred yards from our house near Georgetown, Colorado. And just up the valley, the Eisenhower Tunnel was nearing completion. Mom and Dad understood the impact on our family living so close to an interstate corridor. So, in the summer of 1971, my parents set out to find a quieter and more remote guest ranch in the western part of Colorado.

Crossing the Continental Divide used to conjure up visions of hardships and the frontier. While going onto the Pacific side of the continent no longer held the same mystique, it did have fewer people, and at least where we were moving, there was no interstate highway. It was sort of a 1970s frontier. This adventurous move certainly brought a sense of the unknown, a risk. We were going somewhere new and possibly would discover a reward. My parents' decision to move to Gunnison County, Colorado, seemed to check all the necessary boxes for adventure. So, as you can see, I am a prodigy of adventure. I come by it naturally. I didn't know it then, but I was being bitten by the adventure bug ... and my parents were providing a swarm.

This new home, tucked along Quartz Creek, would prove to be the perfect incubator for young boys in their adventure infancy. Public and private land with mountains, forests, a creek, and just enough free time seemed a good recipe with which to start. The first solo adventure for my brother and me started with the gifts of two bamboo fishing rods. I'm not sure whether they were meant to be so much for fishing as for keeping two boys entertained. The rods were identical—ten feet long with no handles, reels, or eyelets. They were literally sticks. The gift included one roll of eight-pound fishing line, one package of number-six hooks, and strict instructions not to touch Dad's fishing equipment.

At seven and eight years old, we didn't feel gypped in the slightest. No reel, no problem. In retrospect, I now understand part of the parental strategy: identical rods don't get fought over, reels get tangled and hence need adult assistance, eyelets eventually fall off and need repair, and ten-footers can break their tips numerous times and still be functional. I'm not sure whether all this was planned or thought out, but I'll give them the benefit of the doubt, since it occupied two boys for hours on end.

Of course, any good fishing story must involve catching something. Otherwise, it's just a walk by the creek, and there's nothing to tell when you get home. Finding our own worms for bait was a good start to the fishing day. Turning over rocks and sticks seemed to yield enough to warrant the use of an old Folger's can. Then we were off to the creekside.

Anyone who has done much trout fishing knows at times you can easily catch your limit, and at other times you go home skunked. As I recall, our first couple of outings didn't give way to great stories once we returned home. If we did catch one, it was just that. One!

Then one day our luck changed when the fishing gods smiled on us. Upstream from our home was a state hatchery, where they raised trout to a catchable size and then stocked lakes and rivers all over Colorado. The hatchery's plan was to release over one million Chinook salmon, have them go downstream about thirty-five miles to Blue Mesa Reservoir, and then return to spawn several years later. Most of the salmon were in the "fingerling" stage, which put them in the four- to five-inch range. No one told my brother and me about the salmon release or about the closure to fishing during their migration. One other little tidbit that was never explained to us was the idea of a legal limit. I guess my father figured we weren't in any danger of breaking that law. In all our fishing adventures, that day on Quartz Creek will go down as one of the best. Fish size isn't that important to seven- and eight-year-old boys. What is important is quantity. And quantity makes for a better story.

Those ten-foot bamboo rods were seeing some serious action. Since I was the younger (and much smaller) brother, my job was to sit about ten feet behind Matt, who sat creekside. Once I baited the hook, he swung the line over his shoulder and into the hole. Then within a second or two, he swung it back to me. I pulled the fish off and rebaited the hook. Even though the creek ran only twenty yards from our house, we went to a good hole about three hundred yards downstream. Distance is vital in certain activities. Those three hundred yards allowed us to be out of sight and out of earshot. After only an hour of this nonstop action, it was time to gather up the spoils and strut home. All those fish wouldn't fit in the worm can, so we cut a willow branch and threaded them on through their gills. The branch swayed with the weight of tiny salmon. Not only would we never forget that day of fishing; we would also not soon forget the reaction from Dad.

Limit? What is a limit? Size limit? Throw the small ones back? Our dad had lost his mind. *We* now had the story to tell.

We might have eaten the fish, but I really don't remember. Elk and deer meat were the staple around our dinner table. Even though my dad loved to fish, he didn't like eating fish. He preferred red meat, so we had elk and venison most nights. My mom was very creative with the preparation of wild game meat. Dad was also creative in ways to harvest enough meat to feed a family of five and seldom buy beef. He wanted not only to provide for his family but also to help others in need.

On one memorable outing, Dad took my brother and me looking for game to fill our freezer. It was a cold morning with nearly a foot of fresh snow. We piled in our CJ-5 Jeep and started driving some back roads. It wasn't long before Dad spotted some deer. We watched with anticipation as Dad slowly got out of the jeep and took careful aim. The first buck went down immediately as my brother and I looked on with excitement. Dad motioned us to stay put while he took aim at another small buck. Whoa, this was more than we had hoped for. Down went the other two-point mule deer. Finally, as if waiting to exhale, we piled out of the jeep, wanting to shout. But we knew better. Dad had drilled it into us that while hunting, we kept quiet. Now it was time for retrieval.

He instructed us to drag them down to the road. The hillside was steep, and the four-wheel-drive road made a series of switchbacks down the mountain. Dad would drive down on the road just under where the deer lay. Our job was aided by the fifteen inches of snow on the ground and the thirty-degree slope. It really wasn't necessary to drag them, because, with just a little momentum, they slid on their own. These weren't two deer carcasses now; these were warm, hairy toboggans. *Straddle it, hold the head up by the antlers, and hang on.* We were down next to the jeep in no time! Dad could tell we had enjoyed our chore.

One of the deer Dad shot that day was delivered on the way home to a neighbor's house. He was an elderly man, who looked like he was overweight, drank too much, and was very poor. So we concluded that Dad was actually more like a "Robin Hood" hunter, feeding the poor.

So adventure seemed to be waiting for me. Fishing, hunting, and time outdoors beckoned me to jump in with both feet. But I soon realized I was missing something that would prove to be vital if the adventures were to continue. Something, if not recognized and embraced, would leave the adventures empty and hollow. Adventure, I learned, must include purpose if it was to be sustainable. There must be a noble aspect to adventure, and for something to be noble, there must be a higher purpose involved. Some of the adventure stories near and dear to my heart include discovering, having expeditions into the unknown, and charting a course others would follow. That's why I love reading about Lewis and Clark, Shackleton, and the Apollo space program. I was a long way from anything that resembled those expeditions, but nevertheless, they inspired me. It was becoming clear that purpose helped navigate significant adventures.

CHAPTER 2

# North to Alaska

Since I was young, I remember my father talking about his desire to see the last frontier, at least the last frontier in America. I learned a few years later that there were last frontiers on every continent. But international travel wasn't yet on my radar, so the only frontier I was aware of, and the only one my father talked about, was Alaska. Thanks to him, the longing to experience adventures in the northernmost state was planted deep in my soul.

In 1983, I finished high school and enrolled at Western State College (WSC) in Gunnison, Colorado. I would eventually graduate some eight years later. That's right, I managed to pack four years into eight. It's really not difficult to do if you hunt, ski, party, move to Alaska, and attend classes only on occasion. After two years of guiding hunters in the fall and attending WSC during the spring and summer semesters, it was clear that my college career needed a break. A two-year Alaskan hiatus ensued. I clearly wasn't too focused on my studies and had this gnawing in my gut to experience the great north. So I invited my good friend and roommate Kevin to join me. His older brother, Brian, invited himself, and suddenly there were three of us migrating north in May of 1986. We had been told that commercial fishing jobs were waiting for young men like us, and all we needed to do was show up.

I had just turned twenty-one and was ready for some authentic adventure. Sometime in the years just before this, my mother had gone through some noticeable changes. She had started attending a Bible study with some neighbors. This struck me as odd at first, but then I began to notice some differences in her. She seemed content and more at peace than ever before.

I made a mental note. So when she gave me a Bible to take with me to Alaska, I gladly accepted it. I had an old King James Version tucked away somewhere on a bookshelf, but this was different for several reasons. One, because my mother gave it to me. Two, because it was a New Living Translation and much easier to read. And three, because in the front cover she had written, "Spencer, I pray you make this part of your everyday life, because the answer to every situation lies within these pages." Thirty-three years later, I still recall the exact wording and can see her handwriting, though I no longer have that Bible.

So the Bible made the final cut of gear jammed into my 1979 Chevrolet Blazer. The Chevy would be towing an even older Ford pickup, also stacked and packed. We had anticipated Alaska becoming our new home, so we took much of what we owned and deemed necessary. Thankfully I hadn't accumulated too much at twenty-one years of age. It was mostly clothing, camping gear, and fishing and hunting paraphernalia. We each had a few rifles and boxes of ammunition tucked in somewhere too.

Being the confident twenty-something-year-olds, we thought we had a good understanding of Canadian custom laws. No handguns were allowed in Canada. Having obtained this piece of information before departure, we made sure to ship our handguns to the forty-ninth state. What we hadn't anticipated was how Royal Canadian Mounted Police (RCMP) would react when they found several hundred rounds of ammunition for handguns. This was a situation where being young and confident really doesn't help your cause. We couldn't convince the three border guards that we really weren't packing "heat" into their country. What ensued was a vehicle strip search. The stripping, in this case, was thorough and even resulted in the removal of insulation from the underside of car hoods. Several hours later, with our possessions piled on the pavement, we began the tedious process of repacking. With a shrug from the three RCMPs, one turned over his shoulder and without any remorse said, "Enjoy your stay in Canada, aye?"

The idea was to camp and fish our way north, not being worried about where we stopped or for how long. The only destination that was a must was the town of Watson Lake in the Yukon Territory. This small town of less than a thousand is considered the signpost capital of the world. Sometime in the early 1940s, a homesick US Army GI hung a sign with the name of his hometown. The trend continued over the decades, and now there are more than eighty thousand signs hanging in this tiny, remote town. Not wanting to miss the opportunity to represent my hometown metropolis, I borrowed one of the two Ohio City road signs and exported it from Colorado to Canada.

Roughly three weeks after leaving Colorado, we pulled into Anchorage. The first stop was to establish a mailing address at the post office. Next, we headed over to voter registration—not that we were expecting a flurry of letters or couldn't wait to get to a polling booth, but these steps were necessary to establish residency in the state. This took a full year in Alaska and came with some great benefits. First, there was the Alaska oil fund, which entitled every resident to receive about $1,000 per year. But, more importantly, those residents could buy hunting and fishing licenses for about twenty-five dollars instead of thousands. That night we headed to Chilkoot Charlies, a well-established watering hole and spent most of the forty-three dollars we collectively had between us. The next day we were on a floatplane bound for Lake Clark and a firewood cutting job that would last three weeks.

With no expenses and nowhere to spend money, we pocketed our earnings and flew back to Anchorage. I had secured a fishing job prior, so it was goodbye to my mates, who headed for Kodiak Island to find theirs. My commercial salmon fishing was done only two or three days a week, and the rest of my days were spent repairing boat motors and rebuilding a Bombardier Snowcat converted for travel in the Alaskan bush, both of which I knew very little about and still don't. The weeks and months rolled by, and summer was nearing its too-soon end. I had worked six days a week and really hadn't met too many people. I lived on a dry-docked boat in my boss's front yard. The rent was just what I wanted to pay. Nothing! But this was a little lonely, and I was looking for a change.

Another friend from college had told me before I left Colorado that he was moving north too. True to his word, Kurt arrived in Anchorage about four months after me, and we immediately found an apartment and started looking for jobs. His dad owned a used car dealership in Fort Collins, Colorado, so he was somewhat familiar with selling cars. He quickly found a job at a car dealership. We also tried our hands at several other occupational opportunities, like working at a liquor store and selling shoes. The latter quickly became an opportunity to be exposed as a fraud. It seems most mothers buying shoes for their children know much more about podiatry than a twenty-one-year-old adventure seeker. Go figure! And the discount at the liquor store proved to be habit forming. I needed another source of employment. Unlike my roommate, my selling skills were quite limited. But the same car dealership needed a parts delivery person. The main requirements for this job were a driver's license and the know-how to get around town. My one year of auto mechanics in high school plus the three months of self-taught boat motor repair were enough to land me the job. We both now had steady incomes, and it was time to count off the months until we could hunt and fish. Time to save money, study maps, and start dreaming of the adventures to come.

The first few months of life with nine-to-five jobs became somewhat mundane. We could see it becoming darker earlier with each week. It was dark when we left for work, and it was dark when we arrived home. When it wasn't dark, it was cloudy and dreary. This was enough to drive a person to drink … or drink more. We did our part. There was a bar across the street from work, and we were often thirsty at 5:01 p.m. sharp. The bar, while dimly lit, was brighter than it was outside, and with a few drinks under the belt, most everyone watering at this hole seemed to be smiling. "When in Rome." It wasn't long before the occasional trips to the lounge after work became most nights and lasted longer.

"Here, hold my beer and watch this" or "Grab the wheel while I try … !" These are not only punch lines in redneck jokes; they were often slurred by Kurt or me. One night after a few too many, we didn't go home but headed to downtown Anchorage. Now, maybe it's not the biggest city with the most traffic, but when you are from Ohio City, Colorado, with one stop sign, there are a few new challenges to navigating an automobile while inebriated. Those one-way streets proved to be downright confusing. Did I mention you need a driver's license if you are going to be employed as a parts delivery person? Time for some lessons at the DUI school of hard knocks. As it turned out, I was the lucky winner of four nights of free lodging at the rebar motel courtesy of the state of Alaska. The prize package also included a hefty fine, twenty hours of community service, and three months of walking, riding buses, or bumming rides. Soon after my court date, I would also be in jeopardy of losing the job I had taken for granted.

The following morning, I knew what I had to do. So I mustered up the words and told my boss I was thankful for the job he had given me and let him know he would need to find a new delivery person, because in two weeks, I would be without the most important requirement for this job: my driver's license. Some life lessons are better learned with a little pain, and yet other lessons in life seem to be aided best with a dose of loving-kindness. You know, that whole "a spoonful of sugar helps the medicine go down" theory? I was about to learn a lesson that would taste sweet for a long, long time.

Two words are important for every person to understand if he or she wants to enjoy a relationship with God: mercy and grace. My vocabulary didn't even include these words, so to say my understanding of them was limited would be a complete exaggeration. That day in my boss's office, I took my first baby steps toward their understanding as they were demonstrated to me. My boss absorbed the information slowly and thoughtfully. He didn't react or show any emotion at all. In fact, I wasn't sure he had even heard me.

Until this point in my life, I really hadn't given much thought to God. I was pretty sure he existed, and I had a certain reverence for the Bible. This was mainly due to the several "close encounters of the God" kind I'd had. The first one came at Vacation Bible School when I was about twelve. Another came a few years later while taking a friend elk hunting. Each time, someone shared with me about God's unconditional love and the forgiveness he offered. But I had a certain disdain for religion and couldn't yet see that God wasn't interested in religion either. In fact, as one reads the New Testament, it is clear that Jesus came down the hardest on the religious people of his day.

My boss finally looked up at me and thanked me for telling him. He went on to say that he thought I was a hard worker and agreed I had made a big mistake. What he added after that left me with my mouth open and looking like a deer in the headlights. He explained that the parts department needed a person running the back counter, which meant keeping the twelve-plus mechanics in the service department supplied. This would be a promotion and come with a pay raise. Really? Had I heard him correctly? Was this a joke? Did someone have a camera rolling and later this would be on *Candid Camera*?

Now my boss was waiting for my response. I don't remember how I communicated to him that I wanted this new job and the promotion with the pay raise. But I left his office with my head held much higher than when I'd walked in but also with a deep sense of humility. Those two can go hand in hand: no longer ashamed and yet sincerely humbled. I also walked out of his office with the definitions of two words that would later become important in my vocabulary: mercy and grace. What I deserved (and fully expected) that day was to be fired. I was not. I didn't get what I really deserved. That is mercy. Instead, I was given a new and better job *and* more money. I was given something I didn't deserve or earn. That's grace. I now had a working definition for two words that would be vital to me in a few short months. I didn't know it then, but I now had the context for something genuine with God. I started my new job a week later, and my DUI and the consequences of it were never mentioned again. Not forgotten as much as never brought up again.

Chapter 3

# Bush or Bust

We had a few bumps, bruises, and minor liver damage to show for it, but we had made it through our first Alaskan winter. Of course, I had experienced long, cold winters before. I grew up near Gunnison, Colorado, one of the coldest spots in the lower forty-eight. What seemed more difficult to adjust to was the lack of sunshine during those winter months, which may have contributed to the liver damage.

Summer brought many changes for us. Days were longer, and they were warmer and even sunny sometimes. Summer also brought about my one-year anniversary of living in Alaska and with it state residency. Now we could afford those coveted hunting and fishing licenses as residents. We started by fishing for salmon and halibut. We dug razor clams in Cook Inlet and even worked a shift on a dungeness crab boat out of Homer in exchange for the sweet meat. We also left our sketchy apartment and moved to a sketchier fifty-dollars-per-month shack on the hillside above the Anchorage Zoo. The apartment living proved to be expensive and a bit limiting. We couldn't build big bonfires in the parking lot or shoot guns when we wanted, so the move up on the hillside brought a few freedoms we had been denied. There was no running water in our shack, but fifty yards away was a "community" bathhouse and restroom for us and the other shack dwellers living nearby. The inconvenience was worth it as long as we could build fires and fling lead. It was a good trade and a money saver too. For two guys who each spent twenty-five dollars on rent, we sure ate well. The food of kings was often in the bonfire just out the front door or on the single-burner hot plate: crab legs, fried clam strips, clam chowder, salmon, and halibut.

The wildlife in Alaska is amazing. Moose, caribou, grizzlies, Dall sheep, wolves, and wolverines were all on our lists of animals we wanted to harvest. Thanks to penny-pinching and fifty-dollars-a-month rent, we both had nest eggs we could spend on adventures. After several fishing trips, the first fun we chased was a fly-in Dall sheep hunt. We landed on a glacier and set up a base camp. From there, we located and hunted rams. Our pilot seemed proficient enough with his Super Cub equipped with tundra tires. One at a time, he shuttled us from a highway north of Anchorage to a glacier in the Chugach Mountain Range. One week later, we would rendezvous with him and hopefully check off the first animal on our list.

Dall sheep meat is delicious. They are beautiful and amazing animals. I just looked over my shoulder to admire the taxidermy mount in my loft; it came from that first hunt in Alaska. Good taxidermy is really wildlife art, and my wife and I have a lot of art in our home. (I call it "decor by death.") Each time I look at one of these beauties, I can see the place and remember the experience with clarity and detail; even some of the emotion still lingers. This may seem barbaric to some, but it is how I was raised. My wife appreciates it as well but a bit differently. She names the animals and seems to have a relationship with them. Maybe it's a gender thing, or maybe it's her way of dealing with the decor. I'm not sure. I think it's more that she likes that I like them. What a wonderful wife God gave me!

The Dall sheep hunt just whetted our appetites for more. After all, it was only a week-long hunt. What we really wanted was an elongated stay somewhere with daily hunting and trapping. That is what the goal had always been. The dream was to go and live like the mountain men of the past had done while hunting, trapping, and living off the land.

That was the lure of the last frontier: to live somewhere that required grit. This meant being in a place where we couldn't just pack up and leave when it got tough, a place that captured us for a long spell because of the vastness and lack of people, a place that could be endured and enjoyed but never tamed. The wilderness or "bush" as most Alaskans call it is just such a place. But where, how, and when could we go on such an epic outing?

And then as if we had asked the question on the yet-to-be-discovered Internet, the answer came. The owner of the Bear's Den Lodge near Ugashik, Alaska, came into the car dealership where we worked. Not only did Kurt sell him a car, but Kurt also explained our predicament. As luck would have it, the owner of the lodge, which operated primarily in the summer for fishing guests, was hoping to find someone to house-sit in his lodge through the winter. Actually, it was more like the fall, winter, and spring. The lodge was accessed only by floatplane, so by early October, planes quit flying and didn't start up again until all the ice was gone in early May, which meant more like an eight-month stay. Perfect!

Oh, the enthusiasm of youth! We found the spot on a map near the mouth of Lower Ugashik Lake. The Alaskan Peninsula swings south and west out into the Bering Sea, and about halfway down the arm before you hit the Aleutian Islands there is a tiny village on Bristol Bay called Pilot Point. About ten miles upstream on the Ugashik River lies another whopper called Ugashik. After another twenty-six miles upstream, you run into Lower Ugashik Lake and the Bear's Den Lodge.

The owner agreed to fly Kurt and me, along with Kurt's Chesapeake Bay retriever, on a floatplane to the lodge. The dog's registered name was Double Aught Buckshot and was a great companion to both of us. Slightly hardheaded as many Chesapeake's are, he was well trained for bird hunting. The lodge owner also agreed to get all our rations and gear out there as well as several fifty-five-gallon drums of heating oil and several propane cylinders (there were few trees and therefore no firewood on this part of the peninsula; hence the need for heating oil). In exchange, he would have two attentive watchmen keep an eye on his things. His fear, it turned out, was that someone would come up the river and possibly break into his lodge or vandalize it. To his admission, he really didn't know what it was like out there in the winter.

Needless to say, the crime rate around Lower Ugashik Lake in the winter is low. In fact, it's nonexistent, because there are no people. During the eight months we lived there, we saw only one person. We saw him after hiking twenty-six miles on the frozen river down to the small village of Ugashik. It took two days of hiking to get there. The reward was the opportunity to use his radio telephone and call our families for Christmas and the lodge owner to tell him we had finally seen another human being.

Costco wholesale shopping became a hobby before leaving Anchorage. Cases of fruits, vegetables, bags of rice, beans, and flour filled our shopping carts on those last days around civilization. The protein portion of our diet we would have to supply ourselves. Supposedly, caribou and moose were abundant near the lodge. If you lived in those regions and were subsistence hunting, a resident was allowed to harvest four caribou during the winter. Now we're talking! Living off the land, eating what we harvested, and doing all that mountain man stuff were exactly what we had come for.

On September 25 our float plane landed on the Ugashik River and we spent several hours unloading cargo. About 10 days later, the last plane took off and we were alone. The caribou were plentiful. We saw huge herds migrating north out of the Aleutian Mountain Range, and many of them came within a mile or two of the lodge. We learned to sit, wait, watch, and take careful aim. We ate lots of caribou meat just about every way you can imagine it.

Toward the end of our stay, maybe in the seventh month, we ran out of most of our rations. Flour and sugar were gone. Fruits and veggies were gone. During the last six weeks or so, our diet was mostly rice, beans, and meat. Did I mention lots of caribou meat? We had rice for breakfast, beans for lunch, and caribou and beans for dinner. Eventually, creativity drains out of the culinary mind when you have only three ingredients. Our redundant recipe of caribou and beans finally became known as "Caribbean soup." That was the end of the cuisine creativity.

We thought we had rationed our supplies appropriately. We even tried to save a few treats for special occasions like Thanksgiving and Christmas. We might have been in the middle of nowhere, but we fully intended to celebrate with food when we could. Traditional fare, like turkey, was impossible. But there were still some larger waterfowl that hadn't migrated south yet. Swans seemed to be roughly the same size as turkeys. With some luck, we might be able to have a big bird on our dinner table in late November. We found success on our third hunt, and Double Aught retrieved the largest birds of his life. We were set for our Thanksgiving feast, complete with instant mashed potatoes, gravy, and the long-coveted pumpkin pie. We had saved a canned pumpkin pie mix and even managed to make a crust. We had no whipped cream, but since this would be one of our last "real" desserts for the year, it wouldn't matter.

The swan was plucked and stuffed in the oven. Five hours' worth of precious propane was used in the baking of the bird. Careful detail was spent on the pie, which was set outside on a shelf to cool. Once cooked, the trumpeter was carved, and potatoes were served. Cooked swan smells a little different than anything I have ever eaten. It also tastes different than anything I have ever eaten. After one bite each, we were done. The taste was terrible, fishy, and tough. We gobbled down the instant mashed potatoes, and that left us lingering.

"Kurt, what do you say we get that pie and just cut it in half?"

No convincing was required. "All right! Let's eat an entire pumpkin pie for Thanksgiving!"

Salivating again, Kurt cleaned our plates as I threw the swan carcass out the front door for Double Aught to enjoy. The dog had become accustomed to eating whatever we pitched out the door. Usually, it was beaver carcasses after we had skinned them, and sometimes there were leftovers we didn't want. We had brought three extra-large bags of dog food with us. His rations needed to be supplemented too, with caribou meat or other discards. Double Aught usually feasted near the front porch while we ate dinner.

After our Thanksgiving dinner disappointment, we were ready to feast too. I heard a faint clunking noise coming from outside as I went to fetch our cooling dessert. The shelf on the porch was about five feet high, and as I reached in the dark for the pie, I heard the clunking noise again. Turning toward the sound, I could see Double Aught staring at me. As I craned and focused, I could see a light brown ring around Double Aught's nose. Then under his long string of drool, I saw the pie pan and understood the noise. The only thing left was the slobber-soaked crust.

I think I blanked out for a few seconds after that. I was jarred back to reality by Kurt calmly saying, "Spencer, you are killing the dog. Spencer, you are going to kill the dog." I let go of Double Aught's neck and only then realized just how much of a sweet tooth I had and what "hangry" looked like. I also realized that dogs prefer pie over swan too.

Another ration we ran out of sooner than expected was whiskey. Neither of us had ever tried to ration booze before. If we ran out, we went to the liquor store and bought more. So, the question was, how much would we drink and how often would we drink it? Hmmm. Perspective is key here. Either we didn't take enough or drank too much or too often. I remember loading several boxes (cases) of whiskey and other potent libations on the floatplane. Fortunately, we had hit the liquor store lottery a few weeks before our departure. That short employment period at a liquor store yielded something other than just discounts. It also yielded some friendships, which would actually make a donation toward our adventure. The free stuff was booze in damaged containers. Usually that meant a box was torn, or a bottle had been chipped during shipping. But the jackpot for us came when a frequent flier drove his car through the front of that liquor store. Bingo! Lots of chipped bottles, lots of handles on the 750 ml jugs broken, and lots of torn boxes. We were now set for cocktails in the backcountry, just not for as long as we had anticipated. We warmed our bellies for the last time on some cheap scotch around Thanksgiving. Yes, we had saved the worst for last. And the "last" was only two months into our eight-month stay!

There are some advantages to living so far from civilization, but those advantages are apparent only when you are living *in* civilization. When you are actually living out in the bush, they seem harder to recall. What do two fellas do each night when night starts at three thirty in the afternoon and all the booze is gone? We were busy enough during the daylight hours with hunting, running trap lines, and skinning the beavers and fox. And the daily chores of cooking, cleaning dishes, and playing cards could eat up a few more hours. But there are still a few hours left you can't sleep away.

The radio became a close companion. There was one AM station we could get from Dillingham, Alaska, that became rather dear to us. Twice a day, if possible, we tried hard to be within earshot and turn up the volume. One came during the airing of important messages to folks like us, who lived in the bush. The radio program was called the *Bristol Bay Messenger*, and the DJ would read messages that callers to the radio station left. The other time came just after dinner when it was time to relax. That program was called *Radio Reader*, and for an hour or so each evening, we sat and listened to a book being read over those precious airwaves. A little entertainment and a note from a loved one were welcome comforts from the outside world.

One other activity was reading. We had a short stack of Louis L'Amour novels we read and probably reread. And there was one other book that had been packed for this quest, the Bible my mother gave me.

At twenty-three years of age, I had never given a whole lot of thought to death or dying. What kid in his or her twenties ever does? Youth has a way of making some life issues seem distant. Life in the bush, however, closed that distance and brought some of those issues to the forefront. It had a way of shaping me, stretching me, and all the while beating me up. The word *humbling* came to mind. Humility, I later found out, is always a good thing. Humility is the predecessor to teachability, both of which I was badly in need of.

After the honeymoon in the bush was over, some of these lessons began to be impressed on me. There is something about living in a place you cannot leave when you want to that makes a person feel small or at least powerless. I have heard that people who live on small islands often have experienced this feeling. Anyway, I felt small, at least in comparison to my surroundings. I remember thinking this place could eat you up and spit you out in a hurry. The wildness, the cold, the loneliness—these were double-edged swords. I had moved to the bush by choice and wanted to live there and even dreamed of doing so. These were some of the elements that had beckoned me to live in the last frontier, yet now these were the things that could bring a sudden end to the adventure.

The first hair-raising experience that left a mark on my soul came when we were hunting brown bears. You need a big gun when you hunt brown bears, bigger than what I used for hunting elk and deer in Colorado. A friend had loaned us a .300 Winchester Magnum to use while we hunted bears. A .300 like that will do the job if your shot placement is good. The other guns we had brought were smaller calibers for caribou and small game. I had my trusty 30.06, which Dad had given me when I was fourteen. I loved that gun, but it was rather small for bears. Kurt had a .270, which is smaller still and a better gun for deer or antelope.

That September when our floatplane had landed on the Ugashik River near the lodge, we had only half of our gear. The de Havilland Beaver single-engine plane on floats had a payload of around two thousand pounds; it would handle only half of our gear and other necessities. The other planeload was to follow in a day or two.

A day or two turned into three weeks. Unfortunately, most of our ammunition was on the second plane, and bear season opened in a few days. Armed with our undersized rifles and only a handful of ammo, we set out to hunt bears. We weren't about to miss our opportunity to hunt brown bears on the Alaskan Peninsula. I had waited and dreamed of this hunt for years.

We boated across the lake about eight miles from our cabin and set up a spike camp. Several days later, on top of a mountain about three miles from camp, I stood over my bear. I had spotted him from over a mile away. It took me two hours to scale the mountain he was on, only to find him another valley away when I got there. I finally closed the distance just before dark. In my exhausted, excited state, my aim wasn't as true as I had hoped it would be. The bear finally went down, but it took my last bullet to anchor him. Whew!

The next twelve hours turned into a series of mishaps that resulted in the first huge helping of humble pie I would eat in the bush. I was out of everything—food, water, and energy—but I needed to start skinning the bruin. On a steep slope, wedged into a patch of alders, this task proved to be more than I had anticipated. While this wasn't a big brown bear, wrestling the five hundred pounds of dead weight was exhausting. I managed to skin one side and get him rolled over, but the strength and energy needed to skin the other side was gone. Even if I did get it skinned, what was I going to do?

It was dark. My one flashlight was toast, and I had no ammo for my gun. It was early October, so the nights weren't frigid but cold enough. I was getting colder as my sweat dried. I just wanted to rest and get warm. The bear's carcass was still warm, so I decided to curl up next to the bear and drape the skinned portion of the hide over me. Warmth at last. It was now near ten o'clock, and I was able to sleep for a couple of hours.

I woke about midnight, and the first thing I noticed was how badly I now stank. My stench was powerful, but even that was masked by the bear's. I tried to open my eyes as if I had been in one of those deep sleeps, which I hadn't. I rubbed my eyes again, then realized it wasn't my eyes being closed that prevented me from seeing. The problem was, there was nothing to see. Thick, low clouds shaded the tiny bit of light I could have used from the stars or moon. It was pitch black, the kind of dark in which you can't see your hand in front of your face. Immediately, I started shaking from the cold. I had lost my heat source when the carcass cooled. Now what? I may not freeze to death if I stay there, but it would be one miserable night.

I concluded a few minutes later that I was better off trying to get off that mountain in the dark than staying there. I knew this was the lesser of two bad options. I would need both hands free to "feel" my way as I pushed through thick patches of willows and alders. I also knew there was no reason to pack my gun because I had no ammo. I would leave it by the carcass and come back the next day with help. I made my way down a few hundred yards and found a stream I could follow. At least in the stream bed there were fewer branches to deal with. It was slow going—feeling for a branch, hanging on to it, taking another blind step, then repeating the process with the other hand and foot.

After an hour or so, I found the stream becoming a little bigger and steeper. A low rumbling noise also seemed to be getting louder. I stopped several times and listened, but I just couldn't identify the sound. There had been a slight breeze blowing down the mountain all night, which is common as thermals change that time of day.

I made it another ten to fifteen yards over the next five minutes. The rocks in the stream were slick, and I kept slipping and sliding with each step. If not for the branches I was hanging on to, I would have spent most of that time on my butt. The noise now seemed to be close but still not identifiable. I stood there a moment, searching my foggy brain for the answer. Nothing came. Then that downhill breeze suddenly shifted back into my face uphill. With it came a mist of water that hit my face.

The brain fog cleared for a second, and then it hit me. I was heading for a waterfall! I knew then that I couldn't continue down my watery pathway, but leaving the stream meant thicker brush to navigate. I knew what I had to do, but my legs didn't want to cooperate with my brain. After all, it was an exhausted, calorie-deficient, and cold-laden mind that was making the decisions for those legs. Finally, my head won the argument, and my legs agreed to go back uphill and then take a ninety-degree turn into the thick, head-high vegetation. I made it to a tiny clearing and lay down. It took only a few seconds for a few tears to start as I gave into my fears.

"There are no atheists in foxholes or tragedies" was a phrase I had heard before. And even though I had never considered myself an atheist, I hadn't done much praying either. That night in the bush, on a steep slope overlooking Lower Ugashik Lake, I prayed. I prayed like never before. Yes, I was crying out for God to save me, but the prayer came from a place of deep humility and contrition. After about a fifteen-minute prayer and tear session, I made it upright again and restarted my Braille method of bushwhacking. Another thirty minutes of side-hilling gave way to a bigger clearing. At least I thought it was bigger; I still couldn't see. I turned downhill again toward the lake. With more clearings and fewer alders and willows, I was able to pick up the pace a bit.

A little hope bounced into my steps. I was going to make it! The wind shifted back again to the normal downhill breeze, and with it came my brown bear and sweat-drenched stink. Then it hit me; I was walking through bear country and smelling like blood and other foul aromas without a weapon. I told myself, *Push the fear back down! Just keep walking or stumbling, and get closer to camp and my hunting companion.* I finally hit the pebble beach at about three thirty a.m. and knew the branch-face-slapping part of this night at least was over. I still needed to find our spike-camp, a dome tent pitched about twenty yards away from the water's edge. Spotting it would be a challenge. I stumbled another half mile, and to my delight, a dim light was on in the tent. "Thank you, Lord." I exhaled.

To this day, I feel so self-centered when I think about that night. I hadn't once stopped to think about Kurt and what his day and night must have been like. We had parted ways that afternoon. His evening had come and gone without any word from Spencer. He had heard each of my shots just before dark and knew exactly how many rounds of ammo I had, with no word from Spencer. He went to bed that night, not to sleep but to wait and wonder without any word from Spencer. He went to bed that night, wondering whether the rest of this epic eight-month adventure was going to be a solo one for him.

Needless to say, we were elated to see each other. After a few hours of sleep and a good breakfast, I headed back up that steep mountain to retrieve my bear hide. But this time Kurt was there each step of the way. And he had a loaded gun!

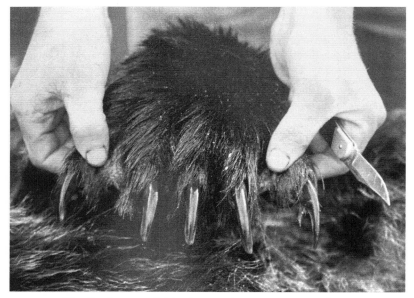

*The final stage of skinning my bear, impressive claws.*

It wasn't long before we had other opportunities to cry out to the Lord again. Falling through the ice is a sensation like nothing I have ever felt. Shock, followed by panic, and hopefully followed by pure exhaustion; I think that would best describe it. We each had a few of those "shock" therapy sessions. Since the Alaskan Peninsula is marshy and boggy where it is flat, there is always water within a few steps. And if you want to hunt or trap during the cold months, you must cross ice on every outing. Eventually, we used fifty feet of rope to tie between us and never walked anywhere without the rope stretched tight. One of us led the way, and the other followed in his footsteps. Who went first? Whoever lost the card game that morning!

I don't think there are any atheists living in the bush either. "Humble yourself and cry out to God" would be a good slogan for life in the wilds of Alaska. It wasn't just the idea that we now felt vulnerable and knew we were no longer invincible; what also haunted me at times was that, if my ticker stopped out here and I found that my time was up, no one would find out for months.

Chapter 4

# Father Spencer?

Now where did I put that Bible? Oh yeah, I hadn't unpacked it yet from the book box. *Sorry, Lord.* I had given higher priority to Louis L'Amour and a couple of hunting magazines, which were indicative of where life's priorities were at that point. I'm not sure that God was even on the list, but the list was about to be rewritten.

Priorities and their accompanying lists are a good mirror of where you are on your journey. They can help pinpoint your exact location on God's GPS. Each time I had one of these "That scared me to death" experiences, it seemed to zoom me in on the map. Each time I zoomed in, I didn't like what I saw. When any of us gets to these clarifying moments in life, we can do one of two things: we can let God show us where he is on the map and how to get there, or we can zoom back out. Zooming in can be painful because it exposes us for who we really are. We were good from afar but far from good! Zooming in gives us the necessary navigation to get where our hearts really need to go. Zooming out may ease the pain momentarily, but it further empties the soul from true fulfillment.

It wasn't just the feeling of being exposed that came with the uh-oh moments; there was also something gnawing at me, something much deeper. I was about four months into living my dream of hunting, trapping, and living off the land. My hair was long, my beard was long, and my clothes had been unwashed for weeks on end. It was all part of the dream (except the clothes part; laundry plummeted down the priority list when we found out how difficult it was to do by hand and how long it took for clothes to dry). Anyway, if you wish for something too long and that wish comes true, often the reality falls short of the anticipation. Well, the reality was falling short. Not all at once but steadily, it was falling. Was this all there was to the dream? It wasn't quite as fulfilling as I had imagined it would be.

I could tell I wasn't alone, since the conversations with Kurt became more about what might be next. In a few months, we would leave the bush, Lord willing, and then what? Both of us loved a National Geographic film called *The Yukon Passage*. Maybe that was the next adventure. We could build a raft like the guys in that film and float the Yukon River. That had possibility. There was a trek to the North Pole too. We talked about that for a time. Neither of us really knew what that entailed, but it sounded like a good next adventure.

It hit me one night in January that I was in a vicious cycle. If I needed another adventure to top the one I was living, then the ceiling was too high. Where would it ever end? I was an adventure junky. It's a good thing the whiskey was gone because I didn't need more than one drug at a time. Adventure adrenaline was the drug of choice now. There was no hangover with this drug. There was something much worse—a great emptiness I couldn't fill, a void that would never be filled no matter how many exciting things I did. A few years later, I would read a quote from Blaise Pascal, a French mathematician and theologian of the 1600s. "Inside the heart of every man lies a God-shaped vacuum." Well put, Blaise. My attempts at self-fulfillment were useless. I was proving his theory correct. I was trying to force the round peg into a square hole, and it just wouldn't fit. Only God fit.

It was the book at the bottom of the box. We had read all the others and a few favorites twice. Only one book was left. The Bible seemed like a daunting task. The few times I had attempted to read it ended with me shoving it aside, feeling like I was reading a textbook on astrophysics.

It just didn't make sense. Pride would then rear its ugly head and huff, "Who needs it anyway?"

This time was different, though. I reread my mom's note inside the front cover. "Spencer, I pray you make this part of your everyday life because the answer to every situation lies within these pages." Had anyone else given me that Bible, I probably would never have brought it with me to Alaska or the bush. Was there anyone on this earth I trusted more than my mom? She was the rock in our family. She was always there for us kids. She had stood with my dad when I wouldn't have. She was the hardest worker I have ever known. Bottom line? I trusted her! She was also the most content person I knew. She had joy that wasn't dependent on her circumstances. I could go on. But I didn't want to. Enough! I trusted her, but the more I contemplated it, it frustrated me. Her life was now in direct contrast to mine. Now I was just angry. How could she be content with working hard on a guest ranch every day, cleaning cabins, and scrubbing toilets? How could she be fulfilled when she never did anything fun, exciting, or adventurous? I, on the other hand, was the one living my dream, but joy and contentment eluded me. What did she have? Why was she so at peace? And of course, the better question was, why wasn't I?

It's good to look in the mirror every so often, even if you don't like what you see. The mirror I was holding up shot back a reflection that seemed empty and at best narcissistic. Reality can be your friend but only if you embrace it. My reality reflected emptiness and a desire for something more. I took my mom's words to heart that day and decided to start reading this book, which had the "answer to every situation."

But where should I start? Thankfully, she had tucked in the Bible a note with a handful of verses and another suggestion: "Start reading the gospel of John in the New Testament. If you want to know what God is like, read about Jesus when He walked the earth." That made sense. But was that what I was really searching for? What God was like? I had the sense I was being pulled into something that had started long ago. I was just showing up late to the party. The more I read John's Gospel, the more intrigued I became. I didn't really have any use for religion.

Never had; probably never will. I saw religion as the cause of so many of the world's problems. It was man made, and it put a horrible burden on people.

The first thing that caught my attention as I read was that Jesus came down the hardest on the religious people of his day. Then, as if a puzzle was slowly being built with each piece in its place, the picture became clearer; something else became evident. God wasn't interested in the religious rules, the dos and don'ts or the show and blow of those who perpetuated them. Instead, he was really interested in me and my heart. God wanted a relationship with me. He wants a relationship with every person who has ever lived. Wow! This realization was kind of mind blowing.

The hurdle that had to be overcome was my sin. I had no problem believing I was a sinner. Denial wasn't a river I was floating down. I began to understand that my sin prevented the relationship both God and I wanted. Another light bulb went on when I began to understand Jesus had paid the penalty for my sin and cleared the way for this relationship with God. The mercy and grace lessons from a few months earlier helped a lot here. I didn't get what I deserved for my sin, and I received something I never could earn. Mercy and grace applied. Another puzzle piece fit perfectly. The bigger picture came a little more into focus. If God wanted a relationship and every relationship starts with an introduction, then I wanted to be introduced. I don't remember the exact words of my prayer that night; the words really aren't important. I was the one who held the last piece of the puzzle. I placed the final one with a heart-felt prayer. To the best of my recollection, my prayer was, "Thank you for letting your Son die for me; and, God, I want what my mom has with you!"

No lightning bolts! Nothing earth shattering or sky splitting. I peacefully drifted off to sleep that night. I woke the next morning … peacefully. As the next few days came and went, they did so peacefully. That was the first thing I noticed. I had peace! The next thing I noticed was an insatiable desire to read that Bible. No more astrophysics; most of it was making sense. Something else new was my desire to connect with God. I wanted to talk with him. I wasn't sure how or where to do that, but I had seen many episodes of *Little House on the Prairie* when I was young. Pa would pray before meals. *Yes, that's what I'll start doing*, I thought, *folding my hands, bowing my head, closing my eyes, and saying grace.* I think that meant saying thank you to God for the food. I could do that.

The first time I suggested we say grace before diving into a plate of caribou meat, I think Kurt asked whether it would make it taste any better. I shrugged and started praying. The prayer was short, very short. The next couple of times, I kept one eye open, partially to be on alert for any flying objects and partially to see what Kurt's reaction would be. Nothing was hurled at me, and he seemed to sincerely be praying with me. I knew that something was changing for Kurt too when one night he had prepared the caribou and was waiting for me to come back into the cabin. He wouldn't start eating until I prayed. Eventually he chided me with, "Okay, Father Spence, pray so we can eat!" I didn't mind my new nickname; I just figured it came with the territory.

Eventually, praying over meals expanded to praying at other times as well. I was enjoying my newfound faith friend, and I somehow knew God was always present and listening too. If I was walking to check traps, I was often praying. When I was skinning that day's catch, I was praying. There weren't many distractions, so the opportunities to commune with God seemed endless. When I look back at those first few months of my relationship with God, part of me longs for that solitude again. It was easy to be still and know that he is God.

Then one day in April, Kurt surprised me with an out-of-the-blue statement. We were walking back to the cabin to get a camera. Earlier that morning I had been out searching for wildlife with the spotting scope. This was a daily activity, especially early in the morning. That morning I had spotted a wolverine about one-quarter of a mile from the cabin. We had seen only two others during our time in the bush, and I had already harvested one. Brown bear and wolverine were the two animals each of us highly sought that year. There were ample opportunities for caribou and moose but not these.

Unfortunately, Kurt didn't find a bear in the fall like I had, so there was a little extra emphasis put on the wolverine. He had made a great shot, and now it was time to properly document the hunt by getting some photos. He suddenly grabbed my arm as we were headed for the cabin. He said he needed to tell me something. "All right, can you tell me as we walk?" He had stopped and didn't take another step. In a serious tone, he told me he had prayed two nights ago before falling asleep. I was shocked but tried not to show it. I asked what he had prayed about. Sort of sheepishly, he said he had asked God something. If God would just give him a chance at a wolverine before leaving the bush next month, he said he would take God more seriously. I asked whether he thought it was a coincidence that he had just harvested a wolverine. "No way," he replied.

And I thought we were bush bound to fulfill a long-awaited adventure. I thought the whole thing was *my* idea. Surely living like a mountain man in the last frontier was so I could chase a dream and check it off my bucket list. But maybe, just maybe, God had orchestrated things so we both would be in a place and time so we might have ears to hear and hearts that were humbled. Maybe God was pursuing us, and he knew exactly what it would take for us to see him. Maybe I was fulfilling a totally different dream, one I didn't know was a dream. Maybe this was God's answer to someone else's prayer.

Chapter 5

# The Return of Two Prodigals

It was a tedious job to pack the seventeen-foot johnboat. Except for food and fuel, most of what had come in on two airplanes would need to leave on one boat. We had lightened the load with the consumption of food, ammo, and clothing. Only a couple of days' worth of food and very little ammunition would make the journey back toward civilization. There wasn't much in the way of clothing left to pack either. That's what happens when you do laundry only once a month. Jeans and shirts were falling apart and were so defiled that fire was the only remedy. That's right, laundry was done with a match once a month. No need for detergent or even a propellent. Grime and grease in clothing ignite quite well. Socks and underwear, however, were given the weekly washing, so the fires weren't nearly as toxic as they could've been. The weight and space vacated by food, fuel, and clothing would now be occupied by pelts and antlers. A few of the bigger sets of caribou racks took up a lot of room, as did the nearly one hundred pelts we were taking back to sell. Beaver and fox skins were the most plentiful, but we also had a number of otters and minks that rounded out the load. Of course, there would be room for two wolverine hides as well.

*Me, Kurt and Double Aught with our winter's catch.*

Mid May found us saying "bon voyage" to the Bear's Den Lodge and navigating the tide-affected Ugashik River. We were going back downstream twenty-six miles to the village and hopefully making arrangements to fly to King Salmon, a sizeable town about one hundred miles north. Once there, we could fly commercially back to Anchorage. What had taken us two days to walk in December took only about five hours by boat and motor. There was a bit more activity in the village than during our December visit.

We had seen only one person during our previous trip, and we were hoping to stay with him again. It took a while for him to befriend us when we met a few months earlier. Roger and his wife had lived in Ugashik for thirty-plus years and had a salmon-jarring business. They put salmon in jars for people who couldn't eat canned foods and then shipped it all over the United States. He slowly warmed up to us on that cold December night, eventually inviting us inside to warm up. As we chatted, some of Roger's crustiness softened as he deemed us harmless and somewhat helpless. He even offered a hot meal and a place to lay our sleeping bags for the night.

Then came the unexpected surprise, the offer of a little screen time. Let's face it, we all love a good movie or TV show, especially if we've been deprived for a while. And on that first trip to Ugashik, Roger asked whether we wanted to watch a movie before bed that night. Our smiles shouted yes before it was audible. He led us to his VHS library of over one hundred movies and said, "Pick one." Trying not to seem overly excited and not wanting our host to feel less important than the cinema soon to come, I continued to engage him in conversation. This allowed Kurt to undistractedly peruse and make the selection for both of us.

Bad mistake! Before I knew it, he had handed our host the videotape, and it was in the player. We made ourselves comfortable and were bubbling with anticipation of the only movie we would watch during those eight months. "What did you choose?" I asked him before the music or title came up. The anticipation I had felt a moment before was quickly replaced by a feeling of being deflated, betrayed, and cheated.

"*North to Alaska* with John Wayne," Kurt proudly announced.

"What? How could you? We are living that! Why on earth would you choose that one out of one hundred?" I berated him, pointing to the stack of videos. "There are movies in there about warm spring breaks on beaches and comedies and even chick flicks! And you choose this?" Now, I love a John Wayne movie, and *North to Alaska* is a classic. But it was too late; Roger was now plunked down comfortably in his chair.

As luck would have it, Roger, our host from five months earlier, wasn't at home as we walked around the village this time. So, we made our way to a few other houses and knocked on doors. Kind, toothless faces greeted us after we knocked on the door of one particularly rough-looking house. Two guys in their forties welcomed us. I feel bad that I don't remember their names now. Years have faded some of those details, but the hospitality they showed hasn't faded at all. The next day they radioed for the airplane to come from Pilot Point, a town on Bristol Bay. Our new friends opened their home, fed us for several days, and even fixed us a much-needed sauna in their bathhouse. I think the latter was a requirement before we could sleep in their house.

The native brothers were enjoyable hosts, and we were grateful for the way they took in total strangers. After three days, our Piper

Cherokee plane was circling the dirt runway in Ugashik. The brothers had helped shuttle all our belongings to the end of the airstrip as we said our goodbyes. Kurt and I wanted to show them our appreciation, so we offered them what little cash we had tucked away from our time in civilization. A hundred bucks wasn't a lot but would convey our gratitude. Their reactions were the same. A shrug and a palm waved us off. They wouldn't think of taking money; besides, there was nowhere to spend it in Ugashik anyway. As Kurt folded the cash back in his pocket, the brothers leaned in, and in a hushed voice one of them sheepishly replied, "Instead, send us whiskey." They leaned back, and those same toothless smiles that had greeted us three days earlier were back. A week later, Kurt might have sent them a package.

The *Duck Dynasty* look wasn't in vogue yet, but with our bushy look, we seemed to blend in with the Alaskans. We were excited about getting back to see friends, eat something other than caribou and beans, and enjoy what civilization had to offer. The problem was, *we* were no longer civilized. Oh, we hadn't forgotten our manners; we just saw the world a little differently than we did eight months before. We sold most of our fur over the next week and loaded up again in that K-5 Blazer, which had been parked for almost a year. Needing more room for cargo, we rented a U-Haul trailer, strapped antlers on top, and started the long drive back down the Al-Can Highway toward family, friends, and futures.

The state of Alaska is different from the other forty-nine states. The people and culture are different—not better or worse, just different. We didn't give much thought to the way we looked there. As I said, we seemed to blend in okay. Not so much back in Colorado. The hair and beards weren't exactly popular in the late '80s, not to mention that we looked quite different than we had a few years earlier. We had lost considerable weight during our time in the bush, changing our appearances even more. We had walked and carried packs for miles daily. We also ate a high protein, low-carb diet, which would make the "keto cult" proud. I think I weighed in at around a buck thirty. We were hairy and lean! Not one to mince words, my brother took one look at me and described me as a cross between Charles Manson and Pee Wee Herman. Flattering, indeed.

The biggest surprise upon arriving back in Ohio City was to find my brother and sister now joining my mom in church and Bible study. And within a few weeks, I decided that was the next logical step in growing my friendship with God. It wasn't easy at first, and I didn't quite feel at home. In fact, some of what I was learning was just hard to hear, mostly because I had seemingly lived life without this relationship with God for the first twenty-three years of life, and now I saw that he had something to say about how I would live in the future. The Bible had something to say about how I spent my time, how I spent my money, and with whom I spent them. With the help of others, I soon came to embrace these biblical ideals, not so much as restrictive and cumbersome, but more like the guard rails going over the mountain pass. They kept me on the road and from plummeting to my own end. They were for my own good.

After this two-and-a-half-year break from college, maybe it was time to give it another try. A good ole college try. I had an inkling I would be a bit more focused this go around—not because I had worked all this adventure stuff out of my system but because the void had been filled. I wasn't seeking to fill the emptiness like before. The thirst for adventure wasn't gone, but I had drunk deeply at a well and found the source of "living water" satisfying. Maybe now I would attend the classes I was paying for and find a real major, one actually listed in the student catalog. I couldn't find hunting, fishing, or skiing anywhere.

My brother had started working for the state of Colorado at the trout hatchery up the road. Yes, it was the same hatchery that had given us that first miraculous catch of fish. Being employed by Colorado Parks and Wildlife, including having all the benefits, was nothing to sneeze at. That didn't sound like such a bad idea. I had always thought that being a game warden would be a good fit for me. And given the fact that I had ceased violating the laws this agency upheld, I thought, *Why not?* I had heard somewhere, "It takes one to catch one." I'm not sure whether that is true, but my experience couldn't hurt, even if I wouldn't be able to put it on a résumé. It was time to re-enroll as a sophomore for the second time and pursue a degree in biology.

College looked quite different this time, mostly because of whom I spent my time with. It seemed that God was pursuing me in this chapter of life too, and that pursuit came from a host of different people: students, a pastor, and my family. Some very dear people would enter my life during this time: people who would lean into me on my journey, people who would pour into me what I didn't even know I needed, people who would become lifelong friends. Fellowship—the real kind, it turned out—was just what I needed. It felt like a greenhouse for spiritual growth.

I soon realized that attending church, a friend's Bible study, and a campus ministry were all part of God's plan to draw me closer. Something else significant happened during those first few months back home in Colorado. My dad, at the age of fifty-two, finally surrendered his life to Christ as well. This was monumental for me. In fact, I couldn't imagine a greater miracle than this. Really! He was one of the hardest men I have ever known. His heart and head were stonelike. Proud, independent, blue collar, and tough were how I saw my dad. Being humble, asking for mercy, and acknowledging his sin were not. But that is exactly the position he was in on August 7, 1988, his fifty-second birthday. After a mini-series from God on getting your attention, he was on his knees next to his bed, and my mother led him in prayer as he surrendered. Like I said, it was a miracle for anyone who really knew my dad. It is a miracle anytime one of us from the human race bows down to our Creator and finds new life, but some nuts are less likely to crack than others. We both had made U-turns and were coming home to our Father. My dad had become my brother and a fellow returning prodigal.

Chapter 6

# Guided by God

The contrast between my first two years of college and my last two was stark. Night and day, dark and light, empty and fulfilling, flunking and dean's list. Needless to say, my life had been transformed, and I knew it. I would later learn that the growth rate isn't the same at each chapter of life, but for now, I felt like I was on some spiritual growth hormone. Those last few years at Western State College would be foundational. I did get that elusive diploma, a major in biology and a minor in chemistry. But the real lasting impact was learning how to walk with God, learning the importance of his Word, and seeing God use me in the lives of others.

What I love about any adventure is the unknown combined with a little risk. Previously, that had to be connected to going somewhere: a place I hadn't yet laid my eyes on, a place that wasn't known to me. I was soon to discover that adventure could be at my fingertips every day. Sure, the desire to see something new is always there, but now I realized it wasn't a necessary part of the equation to culminate into adventure.

My discovery was, in part, that adventure began and ended with God. After all, he is the Creator. Ultimately, it is *his* story, and we all have tiny parts to play. But as the Creator urges and woos us to come close to him, the deeper sense of adventure becomes clearer. Purpose is revealed, and that purpose is directly tied to what's on his heart. What is on the heart of God as you read the Bible? People. That's it. His love language is people. The unknown and risk waiting for us each day lie in the opportunity to share with another person that he or she is on God's heart. God's desire is for that person. "Where is the unknown?" you ask. How will this person respond to you and God? The risk? What if the person rejects me? Worse, what if he or she rejects God? What if I embarrass myself? Been there, done that, going back. What if I don't have the answer to a question the person asks? The whole thing is steeped in adventure.

I have about five or six "little Spencers" running around. They are not the fruit of my loins but the fruit of the Spirit of God using me. I have been honored by friends naming their sons after me, because I simply stepped into the adventure and shared my heart with them. The first time I saw this happen was during that last college stint. Fred was a senior and had become a friend through working together. It was about a three-month process as God drew Fred to himself. I knew God had used me and my story to influence him. I was amazed. How cool is this? I felt like a kid in a candy store with a pocket full of change.

A few years later, my friend and his wife gave the middle name of Spencer to their son. From then on, any time I shared my faith journey and the good news with someone else, I had this sense of being on the edge. I had a tingling deep down somewhere that reminded me of times I had spent alone in the woods, hunting—times when every one of your senses is totally alive and tuned in. There was an adventure to be had anytime, any day, just by entering into another person's life and letting God love on him or her.

I was still headed for what I thought was the destination, a job and career in wildlife management. I was getting some good experience working with Colorado Parks and Wildlife (CPW). I spent two summers working for my brother at the trout hatchery. I also interned through the CPW office in Gunnison for my last three semesters. The internship turned out to be one of the best jobs I have ever had. It paid only minimum wage, but I also gained five credit hours toward my degree while getting my foot in the door. I had the opportunity to help transplant one hundred head of pronghorn antelope from northern Colorado to an area south of Gunnison. My internship involved tracking and mapping this herd for eighteen months. Viewing wildlife is always a good pastime, but when coupled with a paycheck, a CPW pickup truck, and a gas card, it is downright awesome. The kicker came when I was also encouraged to participate in predator control as part of my duties. Thinning out some of the coyotes threatening the antelope fawns was icing on the cake.

During that last and final year (the eighth since I started) of college, I attended a student conference in Estes Park, Colorado. While there, my good friend, John Lamb, challenged me to consider taking a year off after graduating and giving that year to God. At first, I wasn't sure exactly what he meant. Maybe he wasn't aware how prone I was to take years off. Then again, maybe he was. Anyway, he suggested that my best friend from college, Andrew, and I should pray about our next step, maybe even spending a year after graduation to go share some of what we had learned about God. The idea was intriguing to me, especially when he mentioned it would be a great adventure.

The student ministry he worked with, called CRU, had a global reach, which meant the possibilities were worldwide. There were many people in countries who didn't have the opportunities I had to know God and walk with him. Surely, I could take another year before heading down this career path, but this time, the adventure had a purpose greater than I.

It was 1991, and there was a buzz about the collapse of the Soviet Union and the now-open doors for people to travel and live in Russia and former Soviet republics. At the encouragement of a few friends, I started praying for those doors to open and for God to raise laborers for that harvest field. It really was an exciting time, and the more I learned about how people had been denied the good news that had changed my life, the more intrigued I was. There are people all around us every day who haven't heard the message of how God loves them and wants a relationship with them, but there are other people in the world who have *no* opportunity to even hear that message. There are people living in parts of the world where there are no churches; no one who walks with God can tell them the good news, and no Bibles are available to them. These are considered unreached people groups.

One of the Louis L'Amour novels I had read while living in Alaska was *Last of the Breed*. The book was set in eastern Russia, and I was captured by it. The author included maps of this region on the cover pages. On those maps were not only rivers, roads and mountain ranges but also names of the indigenous peoples who lived there. It was the first time I had learned of these groups. I was also captured by the fact that the Siberian part of Russia is kind of like a huge Alaska. Something was stirring deep inside me, and the idea of spending a year in a place like that for a greater purpose now had my attention.

Separating and determining motives can be difficult. Even now, I'm not sure how pure my motives were as I considered this next step. Was it just my desire to have another adventure, or was that desire from God, and was he now using it to direct my future? I'm still unsure. And maybe there is always a mixture of both as we walk with God. It's part me, it's part him, and that's okay. And the part that's me hopefully becomes less, and his part becomes more.

After some prayer, I decided the motives were pure enough to say yes. I hoped there might be a team headed to Siberia. There was some interest from a few others about going to the city of Novosibirsk. The interest didn't last long, though, and soon there was no team left. God had a plan, I was sure; I just didn't know what it was yet. A week or two later, I received a call from my friend John who had challenged us to consider this endeavor. He relayed the information that a couple was trying to put together a team to go into Mongolia and that I should call them.

Mongolia and Siberia both conjure up images of cold, lonely, and "What did you do to get sent there?" questions. Since Mongolia shared a border with Siberia (and also was considered "unreached" by most standards), I thought it would also be a good fit. This country, sandwiched between Russia and China, had been under the control of the Soviet Union and communism for nearly seventy years. With the collapse of the Soviet Union, it was now opening to the West and other countries. The first few missionaries began arriving in 1990. A few had arrived before this, but visas to live there were now easier to acquire, and more folks started showing up in 1990–91. With a population of only 2.2 million in a country that is one-third the size of the United States, it is remote. Of that 2.2 million, roughly seven hundred thousand lived in the capital, Ulaanbaatar. I was intrigued, so I made the call and introduced myself to Warren, a guy about my dad's age.

He wasn't one for much small talk, so the introductions lasted about a minute. Then he asked whether I had any experience working in one of two fields, teaching English as a second language or guiding big-game hunters. He went on to explain the visa situation and that a few foreigners like us might be allowed to live in the country for a year. My answer came with excitement. I doubted anyone would really want me teaching English, but I had guided hunters for four seasons (elk, deer, bear, sheep, moose, and mountain goat). It seemed like it might be a good fit for me to join this team. I was instructed to start raising financial and prayer support, and find a friend who could join me. I had expected the recruitment of finances and prayer partners; I hadn't expected the need to recruit someone else to join me. It's not like you are asking someone to go to Europe.

"Oh, Andrew, where are you?"

By this time, Andrew, my good friend from college and eventual namesake of our son, had headed in another direction. He had agreed to work in the same student ministry but in Oregon. Not being much of a salesman coupled with not having the most desirable of destinations meant tough recruiting. I called the few leads I had been given, and they usually ended with "I'll pray for you!" Then click. "Andrew, are you sure you don't want to swap Oregon for Mongolia?" After much prayer and many conversations with me and others, Andrew chose door number two. Looking back, I still marvel at how God put this plan together. Neither of us had any real mission experience, but what we lacked in expertise, we made up for in youthful enthusiasm. This is proof for all of you who might be tempted to say God can't use you. The longer I walk with God, the more I find that what God really wants is for his children to be available, to say, "Here I am. Send me, Lord."

In the meantime, our soon-to-be director and boss made another trip to Ulaanbaatar to hopefully secure invitations for visas. I was excited to mix this passion for hunting and the outdoors with my new passion for telling people about Christ. My passion bubble was burst when we got the news that we would be getting visas—not as hunting guides but as English teachers. This was a hard pill to swallow, and I began questioning whether this was God's plan. But there was now not only an invitation for a visa from the Ministry of Foreign Affairs with my name on it but also a signed contract to teach English with my name on it. I was committed, seemingly involuntarily, but I was committed. Oh, it was only a year, I surmised.

*I have lived in the bush for a year,* I thought. *I can handle teaching English for a year.* Then we got the news that our students would be from Mongolia State University. "Do they really want *me* teaching English there?" I asked God. "After all, I had been accused of speaking more of a redneck version of the language." A bit of panic started to set in. I couldn't even recall my English classes from high school. Nor could I recall what an adverb was or whether *ain't* was a word. *Oh, what have I gotten myself into? Time to trust.* If God had called me to do this, then this was on him. He would have to provide—and provide daily if I was to wear a coat and tie, carry a briefcase, and become part of academia. *Are we sure there are no visas for hunting guides?* My motives became purer with each step I took toward a year in Mongolia.

*The 'Professors', Andrew and I in Sukhbaatar Square in Ulaanbaatar.*

Becoming a professor at MSU, while comical to those who know me, also proved to be a wonderful experience. It put us in touch with many students and professors who would eventually hear about God. Teaching occupied only some of our time. Our team's main foci were to help set up for the *Jesus Film*, help the fledgling church, and if possible, start a campus ministry. So in September of 1991, we left for a year of adventure, which seemed to be more guided by God than by our choices.

CHAPTER 7

# Goodbye and Sam Bino!

Goodbyes are always tough if it's with people you care about. But when some of those people are supportive of where you are going and what you are doing, the good-byes are easier. That's how my immediate family felt: encouraging and supportive in every way. But there were a few others who weren't quite as convinced that I wasn't half a bubble off. Who volunteers to go to Outer Mongolia? There were probably a few others who wrote this experience off as just another wild adventure. I'm not sure which camp my grandfather was in. Did he think I was nuts, or was it that I was just irresponsible and not ready to settle down? Since my grandfather hadn't yet given his life to Christ, I couldn't expect him to fully understand. But he had an older sister, a missionary in the Central African Republic for nearly fifty years, so there was at least a little context for what I was doing.

I went in for the hug goodbye on his back porch that September day in Denver. My arms' open advance was met by the firm handshake I had become accustomed to. After that he slid his hand on my shoulder and pulled me over for one of his inappropriate comments I had also become accustomed to. "Now don't get one of them pregnant over there. They might cut your head off in that country." It shouldn't have caught me off guard, given the source, but it did. Did he want a response to that statement?

All I could come up with was, "Good tip, Grandpa." Clearly, we saw my reasons for my spending a year halfway around the world quite differently. Then he topped it off by the not-so-quiet advice of "And when you come home, maybe you can get a real job and quit mooching off people." I think my encouragement meter was fully charged after that. Time to get to the airport.

I wasn't very familiar with air travel at that point in my life, and it all felt a little surreal. We checked twelve pieces of baggage, only four of which were ours. We also hauled materials for translation and a host of other things for our teammates. Once on board, we settled in for flights to Las Vegas, then to San Francisco, then to Beijing, and finally to Ulaanbaatar. As we left Denver and headed southwest, I could see our route would take us right over home and Quartz Creek Valley. My mind drifted for a spell as I thought about the people I was flying over right then—family, neighbors, and a few friends I hadn't had a chance to tell goodbye. One of those friends was Brad, whose family's ranch was down the road a few miles. He had been a good friend all through school; we played sports together, hunted together, and shared a lot in common. He knew I was leaving, but I hadn't really told him goodbye. In the seatback in front of me was the pricey, credit-card-activated cell phone. *Why not? It'll probably be the last place I can use a credit card for the next twelve months anyway. I'm calling him.*

Brad's dad had ranched his whole life and had never done much traveling. I'm pretty sure in 1991 he hadn't heard of cell phones either, let alone one from an airplane. His dad answered my call and let me know Brad wasn't home. One other thing his dad wasn't aware of was that I was moving to Mongolia for a year.

I wish I had a recording of this phone call. I started by letting him know I was flying over his ranch at that moment. There was a long pause, long enough for me to start counting the cash I was paying for this call. Finally, he replied, "Oh, you are, are you?" It wasn't a question as much as a statement. And his statement was clear. He was calling me a liar.

Not wanting to take the precious and pricey time to explain the advances in cellular technology, I continued with the next unbelievable statement. I told him I was leaving the country for a year and called to tell Brad goodbye. Another pause came, followed by the exact same words "Oh, you are, are you?" I knew then that he didn't believe a word I was saying. Or he was possibly drawing on some history I had with his son. A few years earlier, Brad and I had a habit of tipping the bottle together, so maybe he thought I was drunk. Either way, his words communicated that he thought I was full of it. I left him with another tidbit. "I am headed to Mongolia, and I'll see you in a year."

I thought he might call me a liar at this point. But after his third pause, there came the same words followed by, "I'll let him know." He hung up.

I knew the call was expensive, but my intentions were good—to tell a friend goodbye. It had accomplished nothing, since his dad was convinced I was either drunk at ten a.m. or just prone to telling whoppers. He deemed it so fictional that he never even told Brad I had called.

We were headed to an ex-communist country, but the only route to fly there was through a current communist country. To make matters worse, China considered Mongolia part of its country, much like Tibet. Several officials at the airport told us as much, and even the Chinese maps we saw made no distinction or border between the two countries. *Hmm. Interesting!*

We had a two-day layover in Beijing, and it happened to be a couple of months after the second anniversary of the Tiananmen Square incident, where a heavy-handed government had shut down protesters in a violent display. It seemed that everyone was on edge and searching for potential protesters. None of this was on our radar at that point. But when we were asked to open boxes containing lots of Christian materials and Bibles, we got a little nervous. Okay, a lot nervous.

The customs official rifling through the boxes spoke no English at all. It turned out he didn't read any English either. He called for another agent to translate for him so he could question us about our luggage.

When asked about what all the materials were, we told him we were English teachers and that this was our curriculum. I was so grateful at that moment that we hadn't gotten visas as hunting guides. Whew! God got us out of that one and seemed to have our backs. This became a theme over the next fourteen months.

I don't think we were being presumptuous with God or rushing into this situation half-cocked, as I tended to do sometimes. There was a genuine sense that God had called us there for this time to do this work. During the months that followed, it seemed we were often in situations that required real faith, the kind that leaves you saying to God, "If you don't come through in some way, we can't continue." It seemed like I was depending on the Lord for things I had taken for granted before, things like being lonely, lacking certain comforts, and being in sticky situations. Oh wait, I think he gave me a bit of training in those situations a few years before in Alaska.

The first word we usually learn in a language that is not our own is usually "hello." Sometimes it's "no" or phrases like "I don't know, "I don't understand," or *"Uno mas cerveza, por favor."* In Mongolian, it was *"Sam Bino."* This greeting is more like "How are you?" It was the first Mongolian I heard a Mongolian speak when we stepped on the Russian-made Tupalev passenger plane in Beijing. The flight attendant greeted us with a smile and said, *"Sam Bino,"* to which we simply smiled back.

There comes a time in every cross-cultural situation when it hits you that you are "not in Kansas anymore, Toto," and things are different. This is especially true for many ethnocentric Americans like I was. I wasn't used to being around people unlike myself or who didn't come from my culture, think like me, or talk like me. I was culturally self-centered. It's one of those "You don't know what you don't know" situations, and slowly, as you live in another culture, you become aware of it. Ultimately, it is healthy and lowers walls between people and cultures.

Once seated, the same flight attendant came by with a tray of hard candy, one per passenger. She seemed more concerned that we each received a treat than that our seatbelts were fastened or our luggage was properly stowed. Once we were airborne, she was up and down the aisle again, this time with small bottles of Chinggis Khan vodka. What? That was a first for me in my limited experience of air travel. Since neither Andrew nor I was prone to downing vodka shots at noon, and since neither of us was a raging alcoholic, we attempted to decline the handouts. Apparently, she had never had anyone decline them before, or our international sign language wasn't sufficient, because we both ended up with bottles on our tray tables. Within seconds we made our first Mongolian friend. Sitting next to us, he shot us a look that needed no translation. He was able to clearly say with a look and a couple of hand gestures that he would gladly take the clear liquid off our hands. We'd just had our first cross-cultural lesson, and judging by his smile, we had passed. That wouldn't be the last time we used vodka to make someone smile.

In the early 1990s, Mongolia, like much of the former Soviet Union, was in economic despair. Everything seemed to be rationed. Most food stores had many empty shelves, and what was stocked on some were only the staples. Flour, sugar, tea, and rice were strictly rationed. Ration cards were given out through employers, and our teaching jobs entitled one for each of us. Each week with the card, we could go to the local market and purchase the basics. Stamps were applied to the ration card, and nothing else could be purchased that week. Most of this made sense to us, even though we had never personally experienced food shortages before. There was one line item on the card, however, that made no sense to us. Apparently, vodka was considered essential for living life in Mongolia. When some of our fellow professors at Mongolia State University found out we had ration cards with no stamps on the vodka line, we made more people smile.

CHAPTER 8

# The Price is Right

Teaching English provided the visas needed to stay in the country. Without that visa, the only option was a tourist visa, which was good for only a month. There was one other benefit to the teaching job. It provided us with an apartment. The building was six stories tall with around 50 two-bedroom apartments in it. These were the typical ugly, gray, concrete buildings prominent throughout Russia and the former Soviet states, which made you think that if you had seen one Soviet city, you had seen them all. Imagination and creativity were not highly valued in Soviet communism. In 1991, most of the foreigners living in the capital all lived in the same apartment building. I think the government thought they could have an easier time keeping a watchful eye on us that way. There was quite the ethnic mix making up our neighborhood—Eastern Europeans, Chinese, and even a few Iranians—but most were Americans, who were there for the same reasons we were. It was a missionary ghetto of sorts, which also provided for some great fellowship we couldn't have otherwise enjoyed had we been spread out across the city. Many of the missionaries who lived in that building became very dear to us.

One of the greatest difficulties about living in another culture is often the food. For many Americans, food isn't sustenance; it's pleasure. That isn't true in many parts of the world. It became apparent that I had often used food to comfort myself when I could. That was no longer possible in Mongolia. The diet for most foreigners changed radically. There are more sheep than people in this country, so the meat that is available and most preferable to Mongolians is sheep. There is a big difference between lamb and mutton. Slaughtering the young from a herd makes little sense when you are raising animals for subsistence. Older ewes and rams beyond reproducing constituted the meat diet for most. Well, meat implies muscle, which is the part I was most familiar with. I was now living in a place where people used the entire animal, including innards and fat along with the head and hide; not one part was wasted. At times, when silverware wasn't available, I was given sheep rib bones to use like chop sticks.

After several months of choking down mutton marbled with heavy fat, I lost weight. Being slender to begin with and facing a very cold winter around the corner, I started looking for ways to supplement our rations. There were other forms of livestock in Mongolia. One could see cattle, camels, horses, and yaks. I wasn't ready to put my palate to the test yet on camels or horses, so we set our sights on beef. Most of the cows we saw, which wasn't often, roamed the city streets and often ate from trash piles. Not really wanting to consume secondhand refuse from dumpster-diving bovine, we began looking outside the city for a possible solution to our dietary needs. Good, grass-fed cows surely were out there somewhere. After we did a little asking around, an English-speaking acquaintance had the answer. "I know a guy" is an important phrase in every language. About fifteen miles from the city was the "guy," and the "guy" had a small herd, mostly used for milking and breeding but never eating. Still, the question had to be asked.

Hard currency is worth a lot in an economy that is in flux. US dollars were about as hard as you could get in those days, and fortunately, we had some cash. There were usually two prices for anything in 1991, the price in Mongolian tugriks and the price in US dollars. When the "guy" heard these two Americans had dollars, his eyes lit up. Doing some quick currency converting in his head, he played his hand. For thirty US dollars, he would throw in delivery. The price was right; in fact, the price was perfect. We could hardly believe we had just become butchers for a yearling, grass-fed, never-in-the-city steer. We even picked out the one we wanted from the herd and watched as it was led into a corral. The animal was then dispatched on the spot. I guess so we could be sure we got the steer we had paid for. Then another cultural hurdle had to be overcome. Why on earth didn't we want the pail of blood or any of the innards? With the help of our friend's translation and me pointing to my small bicep, he understood we wanted only muscle. No heart, lungs, intestines, or fluids were sought. He could keep them in exchange for hanging the meat for a couple of days so the meat could age. He agreed.

It was time to sharpen knives and hopefully put some weight back on. Several days later, a Russian car with a roof rack pulled up in front of the building where all the foreigners lived. On the rack, tied down through the rib cage, was a beautiful sight. Two sides of beef! Under the watchful eye of several of the neighbors, we shouldered the sides and lugged them up two flights of stairs. Nearly a whole day was spent cutting meat and grinding burger. No freezer, no problem. The balcony would become our freezer for the next six months or until the meat was gone. To this day, at the smell of mutton cooking, even lamb, I start looking for grass-fed beef.

We were learning how to survive. Even though much of our time was spent teaching English, our focus was on setting up all the details for the *Jesus Film* work. The *Jesus Film* is the most historically and biblically accurate movie ever made about Christ. It is an incredible evangelistic tool, used all over the world. The film had been translated into Mongolian, and now we were working on translating other materials in addition to managing all the details for the big premiere. It was decided that the first showings of the film would be in the capital and done through Mongolian Films, the state-sanctioned film company. The real reason for its existence was to show Soviet propaganda films, and theaters had been built in every town and village across the country. Now, with no more communist films arriving, these theaters and projectors were sitting idle. The entire country, it seemed, was set up to see this film based on the Gospel of Luke. The first night of the premiere was to be invitation only, with most invitations going to government officials. There was a buzz in the city about the film, in which the actors "spoke" Mongolian with no Russian or subtitles.

Classroom time and detailed work like this left my tank running dry. I needed to refuel by doing something—finding a hobby or just getting out of the city and back into nature. This wasn't an easy thing to do. We didn't have a car or the freedom it brought. We spoke very little of the language, so navigating for this need was difficult. In a country this size, with a population of only 2.2 million, you would think finding a place to unplug and get away shouldn't be that hard, but it was.

One of our neighbors, who had been in-country for over a year, found a way to go horseback riding on occasion and invited us to join her. She had made friends with a family who had about twenty horses and allowed her to ride whenever she wanted. *Catch a taxi and have it take you out of the city. Ride horses for the day and return.* It was a great way to get refreshed. The easy part of this was the taxi. Literally, every car was a taxi. You could raise an arm on any street corner, and a car instantly stopped; even ambulances took a fare if they didn't have someone on the gurney. Horses were plentiful too. The most common mode of transportation outside Ulaanbaatar was a horse. I just needed a way to put it all together so I could go when I wanted.

I had never dreamed that moving to Outer Mongolia and working as a missionary would end with so much involvement with livestock. First, buying a cow and now looking for horses. With a little help in translation and a few more US dollars, the horse part of the puzzle came to fruition. I was able to purchase two ponies, two saddles, and two bridles for one hundred dollars. Once again, the price was right. On Christmas morning, I told Andrew to dress warmly and to hail a taxi. I wanted it to be a surprise. We made our way out of the city about ten miles and had the taxi pull up in front of a yurt. Tied up near the front door were two horses, saddled up and ready. I turned to Andrew and said, "Merry Christmas. Pick one!" He was excited to spend the day riding horses again. He was even more excited when he learned these weren't rentals, but we now owned them. And for just two dollars a month per pony, they would be kept with a herd, fed, and watched over. Again, the price was right. We were now proud equine owners and found the fun and adventure we needed to break up the weeks of hard work. We rode those ponies off and on for nearly a year.

There was one other way that US dollars were a great help that year. After the premiere, the *Jesus Film* ran in theaters in the capital for nearly a month. Advertising posters and billboards were put up across the city. A copy of the Gospel of Luke was included with each ticket. As a result, there weren't too many Mongolians in Ulaanbaatar who hadn't heard of Jesus after this. It was now time to start taking the film to other parts of the country. After one semester of teaching at the university, it became apparent to me and the Mongolia State University that they could do better.

Teaching English just wasn't my cup of tea. They would let me continue to live in the apartment because Andrew would continue to teach. I was now free to start planning trips to take the film to other parts of the country. I now had time and started by taking the film, 35 mm reels, to several places for a couple of weeks at a time. I would also try to set up all the logistics for several other summer teams, which were to come and show the film. We had huge maps of the country with thumbtacks marking towns and villages where the film was to be shown.

The problem with multiple film teams was setting up enough translators and jeeps with drivers to cover nine teams at the same time. These teams were coming from all over the United States and represented a handful of different mission organizations and churches. Slowly, through the network of Mongolians in the church, we found translators. It took longer to find drivers and jeeps that were in good enough condition to handle the rough roads in the steppes and mountainous regions. Much of the population of Mongolia at the time was still nomadic, which meant they lived scattered throughout the vast regions, where they could graze sheep. These film teams needed to be prepared to show the film in an old Soviet theater or show them outdoors using generators and pop-up screens. Some of my favorite memories of film showings came from those days, when several hundred people gathered on the steppes and sat in the grass or on horseback for a two-hour film, a different kind of drive-in.

Finally, as the time for teams to arrive neared, we had many of the pieces now in place, with the exception of one: fuel. The whole country was rationing not only food but also fuel. Eventually, the country quit rationing its supplies and reserved it only for police, emergency services, and postal work. All the preparations were made—films, projectors, generators, translators, and jeeps with drivers—all of which were useless without fuel. Now what? These were the times, as I look back now, when faith was tested. "Lord, you have to do something. Please provide!" I learned that year that God's provision is in his time and often through ways I never would have thought. This way no one could take credit for what he was doing. Living by faith seemed like a way of life during those days. We were constantly in situations that required faith. I long for some of those times, even though they were usually accompanied by discomfort or hardships.

God provided an elderly gentleman, who worked closely with our team. He was retired now from the government, having worked with the Ministry of Foreign Affairs as well as being an ambassador to several countries. Mr. Boyo was still well connected and had the credentials to go into most government buildings and get meetings with high-ranking officials. One morning he told me to put on my coat and tie again (just like when I was teaching), and he would pick me up for a meeting. I asked what the meeting was for and with whom. He simply said, as was his way, "You will see, my son." The next morning found us outside a government building that housed the Ministry of Transportation and Communication. We walked in together after he flashed a government ID at the door. Then he let me know he would translate for me, but I would need to communicate the importance of showing the *Jesus Film*, the number of teams that were due to arrive, and the need for fuel.

The Minister of Transportation was soon across the table from the two of us. The moment was a bit surreal. I kept wondering how I had gotten here. How was it this guy from Ohio City, Colorado, who majored in biology and had a thirst for adventure, now had an audience with an official like this? I tried to be concise and get to the point, but I also wanted to share why this was so important. This man's country and its people had many needs. I explained that Jesus could bring hope and change, something most Mongolians desperately needed.

He listened carefully and understood our predicament. He then called a secretary into his office and told her to type up our request for fuel. She returned in a few minutes with an official-looking document. The minister then took his ink stamp, which represented his office, stamped the document, and signed it. He instructed his secretary to make enough copies for each driver. He then let us know that with this piece of paper signed and stamped, we would be allowed to purchase fuel from any gas station in the entire country without question.

We thanked him after the meeting. He, in turn, thanked me for what we were doing for his country during a difficult time. Like many Mongolians, he said he hoped his country would one day enjoy the fruit of democracy, freedoms, and prosperity the way America did. I think the Lord prompted me right then to share with the minister. I wanted him to understand that the blessings in America he was referring to had come from God. I then pulled out a dollar bill and gave it to him. I pointed to the fine print of "In God We Trust." After Mr. Boyo translated it, I explained that God was the reason for this prosperity. The blessings came from God and him alone. It was a moment I will never forget. Not only did I feel like a seed was planted in the heart of the Minister of Transportation, but we now had the fuel we needed for teams to show the film to many more. Once again, the price was right. I gave one dollar away, and we had the guarantee of fuel for a whole summer.

Chapter 9

# Want to lose some weight?

The first place the *Jesus Film* was ever shown outside of Ulaanbaatar was a town about six hours' drive west. Khujirt was the hometown of two Mongolian men who were among the very first believers in the country. They both wanted the film to be shown there before anywhere else. They had family and friends in Khujirt and wanted them to be the first to have a chance to respond to the gospel. So for two weeks in February of 1992, I traveled to Khujirt with a translator to show the film in a theater. To make things official with the state-run film company and these theaters, I needed to be able to say I worked for Inspirational Films, which wasn't too much of a stretch, since it was sort of a parent company for the *Jesus Film*. I had business cards printed up with my name and title of International Representative. Because I was required to wear a coat and tie (again), some thought I was some sort of Hollywood producer. I tried to explain I worked with only *one* film: Jesus!

Each night we would introduce the film and had a question-and-answer time after the showing. We also took a few minutes to share our stories of how Jesus had changed our lives. It was a powerful time, and the enormity of what I was doing didn't go unnoticed. It is always a privilege to share God's love and forgiveness with another person, but it was truly an honor to be the first to take the gospel to people who had never heard. In some ways the opportunity was overwhelming. I was so moved by the idea that God would allow me this privilege. I loved the chance to share my faith journey with the precious Mongolian people. I loved answering their questions and watching as my friend told them about his newfound faith without translation and in their heart language. I really viewed this as the most important and noble thing I had ever done. This truly was life changing, and I knew it. This adventure was reshaping what I considered important. Even though it was lonely and difficult, film trips like this were opening my heart to some new possibilities, like this is the greatest thing I have ever done. And was I sure that becoming a game warden was what God wanted? Or for that matter, was it what *I* wanted anymore? Cracks were starting to form in the shell of my plans.

Not one of these trips was easy. It is worth noting that every time I took the film to a new part of the country, there were serious difficulties. Nothing ever went as planned, and I was often sick, and conditions were rough. That first trip to Khujirt was the maiden voyage for the film—and a maiden voyage for me in some ways too. It was bitter cold and quite lonely. I had one person around me who spoke some English, but it was tough for us to really share heart-level stuff. He could translate quite well, but when it came to conversing, we were both left wanting more. We both knew that eventually we would need more than what we could give each other. It's amazing to me how much easier it is to do a difficult task if you're doing it with someone else. You can do anything with your friends, with encouragement and camaraderie, but if you don't have those, then you begin to find out just how mentally tough you are. Or aren't.

At first, the showings were packed. Every seat in the musty old theater was taken for the first week or so. It was inspiring to see how Mongolians would sit in a cold, dusty communist building and watch a movie about Jesus. The conditions in our "hotel" room were very similar to the conditions in the theater. (The use of quotation marks here on the word *hotel* should not leave the reader with any mental picture other than an old, cold concrete building with a room. Its two twin beds sagged about eighteen inches, even with my skinny frame on one.) The room temperature hovered around 32 degrees, so to stay warm when I was in our room, I stayed in my sleeping bag.

After six months in-country and despite the beef we were now eating, I had lost considerable weight. Fat isn't always a bad thing; it is a wonderful insulator, and right then I wished I had a lot more. The trip to Khujirt was also a maiden voyage for my stomach as well. I had heard of the love Mongolians had for a drink they considered to be their national beverage. Arak is very hard to describe, and I had been able to avoid this drink until now. It is made and consumed more by the nomadic people, and therefore it is harder to find in the city. I think I would describe fermented horse milk like buttermilk meets vinegar with a fizz and a few lumps. To say this is a taste one must acquire really makes me want to slap someone. Why on God's green earth would I want to acquire that? Arak is made by first milking mares and pouring the milk into big skin bags, which hang on the side of a yurt. Then time does its work, aided by a plunger, which is occasionally thrust into the depths and then pulled back up. There are bacteria in this beverage I guarantee most western digestive systems have never seen.

Sitting in a yurt with a family on the fifth day was my Arak christening. Against my better judgment, I gulped down a bowlful. I'd held out as long as possible and was the last to drink. Everyone else had drunk his or her bowl, and it was now my turn. The cross-cultural peer pressure got to me. Did I mention that I was a people pleaser? I chugged it down, doing my best not to taste it. *Man am I tough!* I thought. *I have an iron stomach. I'm not the pampered American that some are.* I sang this song to myself for about six hours before the tune began to change. Three days came and went without my leaving the sleeping bag except for dashes to the bathroom. Abdominal convulsions are the best description I have for those bellyaches, immediately followed by an evacuation from several orifices all at once.

There was a price to pay, and I hadn't fully counted the cost of these trips. On the one hand, I was doing exactly what I felt God wanted me to do. I loved it. On the other hand, they were some of the toughest tasks ever assigned to me. Somehow, they went hand in hand. I couldn't have the former without the latter and knew it. The honor and privilege of taking the good news to these people were directly linked to hardship. Would I continue? Could I continue? A fetal or squatting position wasn't the place and time to try to answer those questions.

On day four of not leaving the hotel room or its proximity to a toilet, my translator became tired of doing the film showings solo, so he suggested bringing a doctor to see me. Sure, if nothing else I felt cared for, and maybe his bedside manner would bring a little relief. The Mongolian doctor arrived a few hours later, and my friend was back in the translation business. "Mongol medicine is good medicine" was how he began. He looked the part of a doctor from the late 1800s, complete with a small black bag, out of which he pulled a handful of acupuncture needles.

Now, this practice has been around for thousands of years, but it was new to me. Pain is the great doorway through which many things walk. I was ready to try anything, even a needle in my belly, one in each foot, and one in each hand. I'm not sure whether the doctor was smiling because of his great bedside manner or because he had his first patient from America, or maybe he just had a willing patient at all. He proceeded and left me with five long, thin needles waving around in circles for about ten minutes.

The doctor and my translator left the room without a word. I assumed this was part of the treatment. A long ten minutes went by as I uncomfortably watched these tiny flag poles planted in my skin. Just before they reentered the room, it occurred to me to start praying a sanitation prayer over these sharp intruders. The doctor gave me a thumbs-up and restated how good Mongol medicine was. My translator followed up with "How do you feel?"

I paused for a minute. I wanted to feel better. I didn't feel any different at all except for five little pinches. I didn't want to disappoint them and cause their smiles to leave. Did I mention I am a people pleaser?

Not wanting to miss an opportunity for some real medicine and not wanting to be less than truthful, I came clean. I reluctantly answered honestly. "I don't really feel much better," I told them. The smiles slowly faded into concern as the doctor removed the needles. He thought for a moment and turned back toward me. With his fist full of needles, he communicated that I may need a second treatment. He used his index finger to point to places around his face and forehead that made a semicircle. I gulped. Then blurted out that I was suddenly feeling better. So much for truth-telling! I even sat up in bed a little to further convince them. It must have worked, because he shot me the thumbs-up again and said, "See? Mongol medicine is good!" No wonder he thought it was good; no one ever wanted a second treatment.

Slowly the tiny livestock in my guts ran out of stamina. I now understood endurance in a new light. I survived my first battle with Arak. But the ten-round, lightweight bout cost me about twelve pounds, precious weight I could ill afford to lose. We headed back to Ulaanbaatar, knowing that several thousand people had seen the film, heard about our faith journeys, and most importantly heard the gospel. It was time to rest and eat some beef. I was ready for fellowship, food, and fun; maybe a horseback ride was in store. After a week or so, I was back up four pounds, filled with faith, and I could think about the trip to Khujirt with a slight grin. Not a mouthful of teeth smile yet, but my attitude was slowly changing. Maybe I could do another film trip in the near future.

I didn't have to wait to replace all twelve pounds before I was asked to go on another trip. About a month passed, and believe it or not, I had trouble putting the weight back on. There were just very few things to eat that helped with weight gain, and at my age, I still had the metabolism of a hummingbird. Some of the recent memories of the trip to Khujirt had faded or at least blurred enough for me to say yes to another trip. This one would be about a month long. The plan was to fly to the western most province and start there. We would hold a miniature premier for the film in the capitol of Olgiy, a town of about twenty thousand people. We would invite the governor, mayor and other officials and start things off with a banquet in a hotel lobby. I use these terms like *banquet* and *lobby* as I still picture them, which I am fairly certain isn't the way you picture them. But there were tables and chairs, and I did have my trusty coat and tie on again. I never could have imagined that fourteen months in Mongolia would be the only time in my life when I consistently wore out a coat and tie.

We advertised with a homemade banner over the theater for a few days followed by a week's worth of film showings, after which we hired a jeep to drive us to the next provincial capital and did this again. After three premieres in three towns, it was time to fly back to Ulan Bator. This was a trip that confirmed to me I was exactly where God wanted me and doing exactly what he wanted. The privilege again to share in work like this was humbling and honoring. I have no idea how many people responded to the invitation for a relationship with God through Christ on that trip. I may never know, but I knew this was a work worthy of my time and even of the hardships that accompanied it. I also knew this now surpassed other adventures and moved near the top of the list. The work was tough, but I was loving it.

My diet on this month-long outing was different yet again. I steered clear of Arak and drank tea as often as I could. Still no Coke or Pepsi had made the market in Mongolia, but there was a bottled orange soda that wasn't bad when you could find it. The tea in Mongolia is a salted tea with milk. It's better than it sounds, and unlike Arak, one can actually acquire a taste for it. A few other tastes new to my palate were goat heads and groundhogs. There is nothing quite like looking at a platter with a skinned goat head and having it look back at you. The eyeballs were milky from the boiling process, but they still seemed to peer back at me. As the host, I was then expected to use a huge knife to carve off choice pieces for all those at the table. I was instructed to save the best cut for myself. It turns out that the best cut on a goat's head is found at the base of the ear. I found a little meat and mostly cartilage in that area, but I did my best to enjoy it. Now it was time for my thumbs-up and a smile that felt so insincere. I thought for sure everyone could see right through it. Groundhog or marmot isn't nearly as tough to stomach. The meat is a bit greasy, but with a pinch of salt, it can be downed with lots of tea. The hurdle to overcome with this dish is the fact that you are eating a member of the rodent family. It helped to just consider this a culinary adventure.

The real hardship that cropped up on this outing was fear. Fear and faith have a hard time coexisting. Some of the fears started during the first few days in-country. I am blond haired and brown eyed. I am also tall and thin. I didn't blend in with Mongolians; I usually stuck out like a sore thumb. In many cross-cultural situations, this amounts to nothing more than curiosity and novelty. But after seventy years of brutal occupation by Soviets, Mongolians viewed most fair-skinned people with animosity. It wasn't unusual to have a drunk man on a bus get loud and close. Even though I couldn't understand much of what was said, the word *Russki*, meaning a Russian, through gritted teeth needed no interpretation. Some children even carried it a step further by chucking rocks at me as they inaccurately yelled my ethnicity. So I was often on edge in public, especially in the evenings or at night. Vodka flowed more freely as the sun went down, and with it came the courage to express the true and rarely voiced feelings about seventy years of "Russki occupation." As a result, one of the first phrases I learned was "*Bi Americoon*," or "I am American!" This would usually bring a halt to the verbal assault. With children and drunks, the information was slower to be processed and sometimes made little difference.

The town of Ulaangom was the second stop on this month-long film tour and the capital of Uvs province. We were scheduled to be there for a week's worth of showings. There seemed to be more opposition here to the film and particularly to me. The theater manager didn't want to show the film at all. However, our contract with the state film company compelled her to open the theater and provide a projectionist. She made it clear, however, that she would do no more than the minimum requirement and didn't like what she considered "new propaganda." After the first two evenings, we met her again to discuss the times for upcoming showings. I was prepared to show her the contract and thus her obligation again, but when she entered the room, there was something noticeably different about her. Her entire countenance had changed. A smile had replaced the tight-lipped scrunch around her mouth. Then she told us she thought we should consider more showings and even running the film for more days. Of course, we were in favor of this, but I also needed to know what had precipitated this 180-degree change in her.

As it turned out, she had watched the film both previous nights. I thought she had gone home after making sure the theater was ready. I'm not sure where she sat, but she had been at both showings and no longer saw Jesus as Western propaganda. Her life was changed. She had embraced Jesus, not as a film or even a historical figure or just as a good teacher. She had accepted him as her Lord and Savior. She now wanted us to show the film across the entire province in every place there was a theater, and she even offered to organize the showings. The change in her amazed me.

The other source of opposition in Ulaangom came after a full week and left me more shaken than anything else that had happened during the fourteen months in-country. One of my companions on this trip was Miajav. He was the father of the first Mongolian pastor. He was in his early seventies, so he had already surpassed the life expectancy of most Mongolians and was likely the oldest believer in the country. He was a sweet and precious man, who was only about one year old in his faith journey with Jesus. Uvs was the province where he had been born and raised, and he still had many relatives and friends there. He wanted to be the first to take the film there and share his story of how Christ had changed his life. Miajav and I were a team for this week, and we both enjoyed the privilege of what we were doing.

One evening he told me we had been invited to a relative's house for dinner before the film showing. It would be nearly dark for the twenty-minute walk through the muddy streets to their yurt. Mongolians often walk arm in arm or even hand in hand. I had become accustomed to walking this way, especially with Miajav. He often grabbed my left arm as we started off and didn't let go until we arrived. He had a firm grip on my forearm that evening.

We wandered our way through several streets as Miajav searched his memory for the correct one. He led me across a vacant lot with a path that was well used. It was near dinner time, so there were only a few people out on the streets. The pathway was single lane and only about two feet wide, so when you encountered someone coming from the other direction, you often brushed by him or her. A trio of two men and one woman approached us, and it became clear that one of the men was inebriated. The woman helped him and held his left arm.

He was about ten feet from me when he realized I wasn't Mongolian. I saw the look of surprise and the immediate replacement of disgust and anger. By then we were nearly touching. A drunken Mongolian slur is difficult to understand, but he muttered something as he reached out and grabbed my right coat sleeve. I don't think any of us on the path was terribly surprised by this. The woman tugged his left arm, and Miajav tugged mine. Now my arms were spread out like I was the rope in a tug-of-war. Then the drunk man broke free from the woman's grip and reached inside his del.

A del, the traditional Mongolian outer garment, is knee length and tied around the waist with an orange sash. The top of his was unbuttoned to the sash, and often the opening is used like a pouch to carry things. His hand disappeared for a split second. I remember thinking this was a bluff since most Mongolians don't have guns, especially not handguns. No gun but also no bluff. He pulled a hatchet out by the handle and immediately raised it over his head. My reaction was to raise an arm to protect my head, which he seemed to want to split like a ripe melon. But Miajav had a firm hold of my left arm and the drunk had my right. I cringed as the hatchet started to swing down and forward.

It felt like someone hit the pause button right then. I can still see this image clearly, and if I listen hard, I can hear the commotion and even feel the fear rise in me again. I wish I could say I was praying or enveloped with a supernatural peace. I was scared. Period! The only thing I thought about was my personal survival. I didn't close my eyes since I anticipated what was next, because I can replay the next few seconds and often do.

I have never seen an angel that I know of, but that evening in Ulaangom, I experienced what I must call an angelic intervention. The woman's free hand rose quicker than the hatchet and met the handle with swiftness and precision. Unable to bring the blade any closer to me, the drunk lunged a couple of times, trying to bring the hatchet down to his target.

Miajav leaned and pulled with all his strength, and with one big tug from both of us, I was free. We both nearly fell to the ground but quickly regained our balance and placed distance between us and them. Thankfully, no one pursued us. The woman shouted something, either to us or to her companion. We didn't look back; we just got to the next street as quickly as my seventy-year-old brother could go.

I tried not to think about what I might be missing if the hatchet had connected. I tried instead to thank God for watching over Miajav and me. I knew there was a lesson to be learned, but honestly, I lived in fear for a while anytime I was out in public. I slowly became more conscious of the fact that I could do little to avoid what I couldn't see coming. I prayed more often and found myself leaning a little more into God for protection and peace.

About six weeks later, I received a letter in our post box back in Ulaanbaatar. Mail was rare. People tried to send us notes and packages—but not many actually arrived. This was before any e-mail or cell phones. In fact, the first fax machine I ever saw was in Mongolia in 1991. We were still using a Telex machine to communicate with the *Jesus Film* headquarters in California. This letter had managed to weave its way through a host of different countries' postal services and ended in my hands, which was no small miracle. The letter was from a sweet couple in Oklahoma, who encouraged me and loved what we were doing. I knew they prayed for us and the work we were doing with the *Jesus Film*. Their letter to me was dated and had been written on the same day as my encounter with the drunk hatchet man.

The letter specified that they were praying for me. "The Lord himself watches over you! The Lord stands beside you as your protective shade" (Psalm 121:5 NLT). God knew what he was doing, and my faith took a big stride that day. This was definitely not adventure only for the sake of adventure. It was God leading me to step closer to him and further from my comfort zone.

We were incredibly encouraged by the response from this most recent film trip. It was clear that God was opening doors for us and the film to be shown. We felt like the offer to help organize everything to show the film across the Uvs province was from the Lord, and we needed to seize it. So it wasn't long before I was flying on the faith-stretching Mongolian airlines again out to the far northwestern part of the country. Flying in any airplane requires some faith, no matter how many times you've done it or how great the safety record an airline has. Old, worn-out Soviet planes made up the fleet for MIAT, the state-run airlines, which offered the only domestic flights in the country. And since driving would take more than a week, it was time to fly. Seatbelts were encouraged but not mandatory, and stowing baggage was viewed as silly. After all, your lap can hold a lot of precious cargo, so use it.

There are several things you hope to never see when you get on an airplane. Rust is one. Another is the cords fraying out of the bald tires under you. I often like to say that "reality is your friend." But in this situation, I preferred to know and see as little as possible. Just close my eyes and pray. That's probably the best way to start another adventure.

Chapter 10

# The world is round and has IVs

Eventually, I opened my eyes during that flight. The bleating of a sheep in the center aisle brought me back from my prayerful state. That was a first. Clothed in a burlap sack with only her head sticking out, the ewe seemed about as calm as I was about flying. I wanted to whisper, "Just close your eyes and pray," but if I started talking to a sheep in a burlap sack, which was lying in the aisle of an airplane, someone might put me in burlap too or a straitjacket. Best to ignore the nervous sheep and keep praying.

This trip would be a little different from previous ones I had made. It was now early summer, which made everything easier, more pleasant, and even beautiful. The Mongolian steppe in bloom is really something to see. I liken it to parts of Montana but with no paved roads or fences. Wave after wave of grass-covered hills, void of people for miles, gives you the feeling of being on a prairie ocean. This trip was also different because another missionary joined me along with a Mongolian pastor.

Martha had been in the country about six months and worked closely with our team. She is a treasure and one of my faith heroines. We also wanted Martha to connect with our hostess and organizer of this film trip. Oyun, the theater manager, had expressed her desire to become a follower of Christ during my previous trip, and connecting with Martha would help as she started her faith journey. It was also nice for me to have another native English speaker along and would make for great fellowship. I also enjoyed watching God knit the hearts of these American and Mongolian women together as sisters.

Thanks to Oyun, all the organizing was done, preparations had been made, a jeep and driver had been hired, and we were off to show the film in some of the county seats that had old theaters. Every province in Mongolia is further divided into counties, and each of those, at least at one time, had movie theaters. Some of these county seats might have only a few permanent structures since most people live in yurts. One structure is always the Soviet-built theater. It always took some time to find the right person with keys to the theater, then make sure it still had electricity and then get it cleaned. Many of these buildings hadn't been opened or had a film shown in them for a decade or more.

The other antiques we became familiar with were the old Russian jeeps. Japanese cars had yet to make their appearance in Mongolia, so everything on the roads was Russian made. Since Russia was producing few cars due to their economy, the vehicles in a Soviet satellite country like Mongolia were the hand-me-downs. Rust and rotted tires seemed like a transportation theme. These UAZ jeeps were the same design as the old Willey's Jeeps used in World War II. The American automotive industry had replaced these four-cylinder, square boxes; Russia had not. Needless to say, we were often broken down, looking for parts, or pushing one to pop the clutch to get it started. Yes, it all added to the adventure, but the novelty wore off quickly. Eventually, antiques need to retire to museums or junkyards.

One mystery I never was able to solve on this trip was the complete disappearance of roads. Now, to call them roads is a bit misleading. A two-lane cow path would paint a more accurate picture. There were places where the path was clear, and direction was obvious, but after cresting one hill—poof—it would vanish. We spent many hours backtracking to search for routes, a tire track, or some other clue as to how to proceed. A compass became quite useful during that journey. On rare occasions, we ran across a shepherd to ask for directions. A big flock of sheep led us to the local knowledge that would often end with directions based only on landmarks. You know, the old "turn left at the fork in that valley and don't miss the grassy hill with a rock" kind of directions. That trip confirmed something to me and also led to a discovery. Some gender-based concepts don't have cultural barriers. These seem to be universal truths about males. Men from any culture don't like to ask for directions. Furthermore, men also love to be asked for directions and give them, even if those directions are completely inaccurate. The perception of knowledge is intoxicating, and most of us males love it. Eventually, to expedite the journey, we came to a consensus: don't stop, don't ask, pray, and keep going in a logical direction.

One such stop led to directionally challenged advice that was so spectacularly bad, it took us in a huge circle. This stop also provided some of the much-needed humor that would leave us chuckling for weeks. We were on about our third game of "find the road" for that day, so we headed to the nearest rise and started scanning again. A large flock was spotted, which meant it would have at least one or two caretakers. As we pulled up, the two young men dismounted from their horses and walked over toward the jeep. Greetings were exchanged, and then everyone assumed the posture that accompanied small talk on the Mongolian steppe—butt to the heals with knees spread out and arms between. It was a position I found painful and nearly impossible to get out of. Mongolians loved it and that made me think they must have another joint somewhere that my genetic code lacked. As they all squatted, I plunked my butt down in the grass and leaned back on one elbow, making sure to avoid any sheep excrement. After our new inaccurate travel target was acquired, they all chatted for a spell. These young men were about fifteen years old and already showed the signs of weathering from hard lives. Continual exposure to wind, sun, and cold while on horseback hit the fast-forward button on their aging. They looked like they could be twenty-five to thirty.

The pastor, Bold (his name, not a description) the son of Miajav, was fun loving and playful. He loved a good joke and any reason to laugh. He had made it a practice on this trip to ask people where they thought I was from. Mongolia as a country is one of the most remote on the planet, and we were in a very remote part of it. The answer never varied about my ethnicity; they all said, "Russki." Bold loved the response he got after the curtain was lifted and he revealed me as an American. I was now accustomed to this his little game and always played along, sure to never spoil the surprise. Even with my limited language, I could follow this conversation.

Bold asked his question while pointing at me. Nothing bursts the bubble of a comedian quite like when the punch line gets no laughs. The young men didn't show any signs of surprise or excitement. Bold just squatted there, unsure what to do. In the past, the responses varied, but he had always managed to get a "WOW" look, no matter who his audience. These two prematurely aged shepherds gave nothing—not a nod, shoulder shrug, or even furrowed brow. Nothing!

A few seconds passed, and Bold started the tedious job of unpacking his anecdote. I saw him using his arms to make the shape of a sphere and then recognized the word for *ball*. *Oh no!* I then started to wonder just where we were. The professor was letting his students know the truth about the shape of the earth, that we are on this side of the big ball now, and on the other side was where Spencer was from.

Finally, a response. The students quickly let the professor know this idea of a round planet wasn't news to them. Okay, we had agreement that the earth wasn't flat. A few more minutes passed as Bold began to wonder whether he hadn't been clear about something or had withheld part of the information. Then one of them asked a question that left those who understood Mongolian laughing hysterically. It took a few more minutes before Bold could translate for me. One of them had asked why a Russian was coming from the other side of the world. In his mind, the only Caucasian faces in the world belonged to Russians, who lived just to the north. So, why would a Russian come from a country on the other side of the ball? We all enjoyed a laugh and often recalled this story over the next few weeks. But it also made me more aware of just how far I had traveled. I let this truth sink in and reflected on my journey over the last few years. This was a story of adventure only God could author.

In most of these county seats, I met with town officials. The meetings were largely ceremonial, but we felt like it was important not only to invite them to the first film showing but also to ask for their permission, even though we didn't really need it. The gesture went a long way and showed our desire to honor them. As is often the case, showing honor is usually a two-way street. Some of these men and women felt the need to show me honor too. One of the most common ways to do this was a gift presented in public. Part of the reason for showing me honor was just for show and reciprocity, but often it was attached to the fact that I was the first American to ever come to their town. The gift was usually the same, even though there was no protocol to follow—an old Soviet-style medal with the name of the town or province on it. These were presented and pinned on my shirt or jacket. Only once did I pin all the medals I had received on my jacket at the same time. I looked like a veteran who had served a lifetime in the army.

We had zigzagged across the province, showing the film each night and giving out the Gospel of Luke to attendees. We were now headed back toward the provincial capital with about five towns out of sixteen still to go. This trip had its challenges, but none were too terribly difficult. We were only a day or two behind schedule, which meant we were really way ahead of schedule. After two weeks on the road, we were all tired, but everyone was in good spirits and healthy. Often during these trips, we never had any idea where we would spend the night. If there was a "hotel" in a town, we usually got a couple of rooms. Many nights were spent on floors of yurts with a family we had just met or the home of the theater manager. There were five of us on this trip, so to feed five people each day was always part of the faith venture. Sometimes we offered to buy a sheep from a family if they could prepare it and feed our team. We were back in the livestock business, except I didn't get to choose which old, worn-out ewe would be bleating in the jeep on the way back to be slaughtered. I wanted to try a lamb, but that was a cultural hurdle I wasn't getting a Mongolian over.

I won't give details of the dispatch and slaughter, but suffice it to say, this biology major learned some good anatomy lessons as I watched. The next step was processing the mutton, which was a family affair. At home, it is referred to as "meat processing," and meat means muscle. Here, it was entire animal processing. First, the intestines were pulled out and cut on one end, tied into a knot, and then stripped to relieve the lining of all the unwanteds. Next, that same lining was stretched open on the end, and the blood saved in a pail was ladled in. Once the intestinal tube was filled, the other end was tied off. The two-inch-by-two-foot log was set on the yurt's roof to coagulate and cure in the summer sun. Yum! Blood sausage.

Thankfully, because we stayed only one night with each host, the sausage was never cured or served while we were there. Cooked fat and meat were often served for dinner. "Sandwiches" were often handed out within an hour or two. Two thick slices of marbled fat were used instead of bread with a slab of cooked liver in the middle. No wonder I had a heart attack at age fifty-two! There just weren't many menu items on these trips that made me want to eat. Another by-product set on the yurt roof to cure was cheese made from sheep milk. This had the appearance of white block cheese but was about a five on the hardness scale. It was tooth-breaking hard and had to be sucked on like hard candy before chewing. Somewhere along this unpasteurized journey, my delicate digestive system gave up. The trip had been going so well until that point.

Once again, some form of tiny livestock took over my body like aliens. I never saw the theater in the next town. The only building other than our "hotel" I entered was the outhouse. We stayed in this town for two extra days, mainly because I was unable to travel. A love/hate relationship is the best way to describe how I felt about the outhouse behind our hotel. I loved the proximity and never wanted it to be too far away, always within short walking distances of no more than twenty yards.

The hate part came when I went through the door. "Squatty potty" is what we dubbed them. There was no seat, only a hole in the floor and maybe a raised platform for each foot so as not to be on the same nasty level as the floor. It seemed that accuracy wasn't a highly held value, so foot placement was of vital importance. Okay, it was more of a hate relationship. On the third day, I no longer had the strength required to get into or out of the needed squatting stance. In fact, I no longer had the strength to even attempt the twenty yards. I was in bad shape, dehydrated, unable to walk, and not completely coherent. Martha and Bold loaded me up for the drive back to the provincial capital, where there was a "hospital."

Quotation marks are important here again. Remove from your mind every picture of a hospital and replace it with an old, worn-out, two-story building with a handful of rooms featuring cracked plastered walls and peeling paint. My room was on the second floor. With a lot of help, they managed to get me upstairs and into a bed. Several nurses and one doctor began checking me out. The diagnosis was dysentery, dehydration, and doubtful. I badly needed fluids, and I wasn't likely to finish the remaining film showings. They did have IV bags, however, with precious saline solution my body thirsted for.

To this day, I still have a hard time comprehending that tiny "hospital" in northwestern Mongolia having IV solution. Thank you, Lord! Nothing that passed my lips stayed in my body for more than a few minutes; even the anti-diarrhea Imodium tablets seemed to exit my lower GI tract with more force than they entered. So IVs saved my life. I spent two days chugging the solution through my left arm. I didn't get up once during those first two days. Finally, after many rounds of liquid in a bag, I was rehydrated enough to need to use the bathroom. I made my way with help to the bathroom on the second floor, tugging my IV stand with me into the small, one-stall restroom. I was overjoyed by the small amount of success.

I turned to the sink to wash my hands but stopped in my tracks. The sink was small and had a shelf on the right side with a drying rack on it. Both the sink and the rack had several dozen IV needles in them, either soaking or drying. Maybe the staff cleaned them before throwing them away. I really wasn't sure. Two things I was sure of, though; it was time to pray the sanitation prayer again and check myself out of this hospital.

I convinced my nurse that I was well enough to be discharged before more IVs were administered. I was incredibly thankful for the fluids, without which I'm not sure I would have made it. But I could no longer stay in this hospital. I was strong enough to have my faith muscle stretched again and get on an MIAT flight back to Ulaanbaatar.

My faith muscle continued to get exercise though. I realized there were two parts to this exercise routine. One was voluntary and involved trusting God for things I wanted. The other was more like physical therapy. You know, where someone is telling you—dare I say, forcing you—to do some exercises. Ultimately, they are for your good, and you will be better off if you do them, but you may not enjoy them at the time. My faith exercises involved trusting God for something I wished I hadn't been a part of, but leaning into the Lord was critical for both the attitude and the outcome.

Since my body was still ridding itself of Imodium tablets as quickly as I could swallow them, I knew I needed something more. Sure, Mongol medicine is good, but I also knew there was a Russian hospital still staffed by a few Russian doctors and nurses. It was a sliding health care scale I was now on. Russian health care was only a few decades behind most Western medicine, which made it about half a century ahead of what was available in Mongolia. Once back in Ulaanbaatar, I was off to the Russian hospital. It was near our apartment, so I enjoyed visits from teammates and other concerned missionaries. Several days later, after a host of powdered medicines I presumed were antibiotics, I headed to rest up in our apartment. I was now down to about 125 pounds. Bantam weight isn't a good weight if you are over six feet tall.

I may have been physically miserable at times, but I was spiritually healthier than ever. There is something about being forced to trust God that is crucial for our growth. It's a vital ingredient I had lacked, in part, before coming to Mongolia. It was around this time that I had another realization. This one was slower to come but much more poignant. I had left the United States nearly a year ago at this point, and at the time, my future had been clear—or so I had thought. I would give this stint of fourteen months to serving the Lord and the people of Mongolia, then return to what I wanted to do: pursue a career in wildlife management. The chorus line of one of my favorite hymns, written in 1922 by Helen Howarth Lemmel is,

> Turn your eyes upon Jesus.
> Look full in His wonderful face.
> And the things of earth will grow strangely dim,
> In the light of His glory and grace.

The realization was that I could no longer remember the attraction to a career as a game warden. It had grown strangely dim. At the same time, it occurred to me that I was doing exactly what I wanted to do. I remember one day telling the Lord that I would do this for the rest of my life if he would let me. Of course, I later understood I could live this same adventure with the Lord no matter where I was or no matter my profession. The question I needed to answer was, where is his adventure for *me*? If we are all unique individuals with unique gifts and skills and talents and passions and desires, then he has unique places for each of us to serve. I knew then and there that my plan for a career in wildlife management was just that: *mine*. I had never asked God what his plan for me was. I was now in a unique position to ask and listen to him. So, during the summer of 1992 in Outer Mongolia, God opened the next door in this adventure for me to walk through.

CHAPTER 11

# My Second

Had God called me to go into vocational ministry? A few people, including myself, asked me this. To be honest, I wasn't sure what that meant. Had God called me? Did I receive a calling from God to become a missionary? I'm still not sure. I wrestled with this for a while. Did I need a special calling before saying yes to join staff with CRU? Why did I need a calling for full-time ministry and not for any other occupation?

What I was sure of was that God had called me to himself and commanded me to share my story with others. Also, I still had this drawing or hunger to go to Siberia, and I didn't see a line of people waiting to go there. I knew some of my motives were still less than pure—adventure, hunting, fishing, and so forth. But in the moments of contrition and humility with the Lord, the pull to Siberia seemed the strongest. There were people north of the border who were waiting to hear about God too. There were entire people groups who had been deprived of the opportunity to respond to God's gift of love and forgiveness.

Was God telling me to go there and tell them? I had been at the border between Mongolia and Russia several times. Each time I was close I did something if no one was watching. This is where you might roll your theological eyes at me or at least do some theological tail sniffing to make sure I'm not whacked out. I picked up a rock, prayed over it for the people who lived close, and chucked it over the border fence. I wanted my prayers to touch that land in some way. And if I couldn't stand on it, then I figured I'd throw a rock onto it.

Moving forward, I felt confident about two things. I would join full-time staff with CRU, and somewhere in the future, God wanted me to head to Siberia. So, after fourteen months, with much more clarity and much less weight, I left Mongolia and returned to the United States. I knew the guy leaving was a much different person than the guy who had come to this country. While I was sad to leave people and a place that had impacted me so much, I was also bubbling with excitement about the next few chapters on this adventure of walking with God. There were many unknowns, but adventure must have those. God had shown me just enough to take another step, which is what walking by faith is all about. I was looking a few years into the future and asking God what I needed to see this vision become reality.

There was a handful of items on a list I thought were important preparations before moving to Siberia. The first was a team. I had seen the great value of being part of a team and had spent enough time alone while showing the *Jesus Film* over the last year to convince me of this. The second was an airplane. Siberia is like a huge Alaska, where the use of small aircraft is essential. There are few roads and only one railroad across the vast eastern part of Russia. The third was a wife. I longed to be married and start a family, and I often prayed that God would bless me in that way.

If someone told me about reverse culture shock, I don't remember it. I remember receiving advice about moving to another country and being informed of things to expect but not about moving back to my home country. Why would I need advice about moving home? After all, it was home—you know, where the heart is—with family and familiarity. How hard could it be? But something changes in our psyche after living in another culture, especially if that culture is quite different from that of your home.

After struggling to find food I actually wanted to eat, there was now a plethora of options. Going to the grocery store was overwhelming. I remember trying to just buy a loaf of bread and not being able to make a decision. In Mongolia, the only decision was to buy or not to buy. When bread was available, there was only one kind, so if your diet needed some balancing, you bought a loaf. I concluded after some time in Mongolia that there were only two food groups. I had been taught in school there were four basic food groups, maybe five. That's how people in first-world countries talk. In much of the world, there isn't the luxury of food groups. There are only two, movers and blockers, and the dietary goal is to balance those two.

Anyway, I was now back in a first-world country and in front of shelves and shelves full of food, and purchasing didn't require a ration card. I really didn't know what to do. I think I came out of the store that day with a pack of chewing gum. I'm not really sure why; I just panicked and thought I had better buy something.

Another shock that came with moving back to the States was how people acted around me once they found out I was a missionary. Some were indifferent, some were excited, and more than a few were shocked. The shocked ones were people I had known from high school or college. They knew me in my BC (before Christ) days, remembered me as a partier, and just couldn't wrap their heads around that much change. I had the chance to attend my ten-year high school reunion. By the time I made it to the picnic, word had already spread. "Nicholl is a missionary!"

"What? No way!"

The few words shared with me by one friend actually spoke volumes. "Well, good for you. Good for you!" He had no idea what to do with me. There were a few other people who meant well but were just clueless about where I had been or what I had experienced. In one church, when a sweet lady asked how I liked my time in Africa, I turned away slightly and under my breath answered to no one in particular, "Not as hot as you might think!" Oh well, I couldn't expect everyone to understand where I had been or even the geography.

As the weeks and months went by, the shock of transition slowly wore off, and I adjusted to life back in the States. I also began to fill out my jeans again with the help of fast food. Not long after getting my legs under me and feeling healthy, I began working on my private pilot's license. I had really enjoyed the time I spent in small aircraft in Alaska. Flying in smaller airplanes sort of gets in your blood and is hard to get out. The main motivation in flying, at least what I told myself, was for missionary work in Siberia. It was about a three-month process, but by late spring 1993, I had passed my tests and was starting to build up some hours.

I had also gone through the new staff training with CRU and would soon report to Colorado State University. I was really looking forward to being on a campus ministry team, sharing my experiences with students, and flying whenever possible. I was also undeterred in my quest for a "P-31." That isn't an airplane. Proverbs 31 describes a wife of noble character, and I was happy to be back in a place where I was around more P-31s than I had been in for a while.

I met Annette a few months later at a student conference in Estes Park, Colorado. She had been on CRU staff for about eight years and recently moved to the University of Colorado at Boulder. We were paired up to work together that weekend on something. I really don't remember what we were supposed to do or what the work was. Needless to say, I was a bit taken. She was beautiful, close to my age, and serious about sharing her life with others. And she had already spent a year in Russia (1989–1990). *He shoots. He scores!* That's how I felt, even though I was trying to play it cool and not like I had just come back from fourteen months in Outer Mongolia.

I had asked myself for some time what the chances might be of meeting someone I really liked who was willing to go to Siberia. This girl had moved to Russia and worked with students when it was still the Soviet Union. She had lived in Leningrad, now St. Petersburg. This girl was *all in*! Soon I was meeting other women and students whose lives had been impacted by Annette. It seemed like once a week I met or heard of someone she had led to Christ, trained, or discipled. She didn't have to say she was *all in*; it was clear from the path she had walked.

Now I wanted to be *all in*, with her. We shared a lot about our experiences in Russia and Mongolia, and we really enjoyed hearing each other's tales of life overseas. I wondered what I had to offer a gal who had already done so much and been so far. Then it hit me! *See if she wants to go to Mongolia for the summer and show the* Jesus Film. We could recruit some staff and students and make a summer of it. I wasn't surprised when she agreed. She would colead the summer team with me. My motives were again mixed.

Working together was one thing, but I wanted to date this gal. I had gotten to know her and had seen who she was, what her life was about; and as I said, I was all in. But we hadn't even been on a date yet. I was ready to try to sweep her off her feet, so I rented a single-engine plane, and we flew to Steamboat Springs for lunch and then back to Fort Collins for dinner. It was a magical day and a great first date. The more time we spent together, the more I knew she was the one. She was a bona fide P-31 and had stolen my heart.

After eight months, two of which were in the romance-rich environs of Mongolia, I asked her to marry me. Of all the gifts God has given me, Annette is second only to God's gift of his Son. We both felt the same, so we began signing all our notes and cards the same way. Nearly every note, card, and e-mail for twenty-seven years has been signed "Love, your Second." It's a reminder to each other that our first love is always God, and we are each other's seconds. I would soon find out that God was leading me into the next great adventure—the mystery and joy of becoming one with Annette. Walking with God, side by side with Annette, has been the greatest joy of my life.

Chapter 12

# Siberia and a baby

We were married on January 7, 1995, surrounded by family and friends. We wanted our first year of marriage to be in the United States. We just figured it was best not to put the extra stress of moving and living in another culture on our young marriage for at least a year. That would give us a little time to get to know each other better, work on the same campus team, and start life in a small apartment near Boulder.

We both look back at that first year and smile. The apartment we could afford was a one- bedroom loft next to the railroad tracks about five miles out of Boulder. Being that close to the train at an intersection gave us newlyweds the opportunities to work through some differences we might have, such as, how we both handle the train whistle at three a.m. when it sounds like it's coming through the apartment. Or how many of Spencer's twelve taxidermy mounts (wildlife art) would be part of the decor in a one-bedroom apartment? I am still amazed that Annette generously agreed to five. City girl goes country!

Yes, we had a few geographical hurdles to get over after getting married. Annette was a Midwest city girl, and I was a couple of degrees away from full-blown hillbilly. This became very apparent one day about three weeks after the wedding when Annette asked what their names were, referring to the taxidermy mounts. Caught a little off guard, I said I had never thought about naming them. This led to two important revelations: First, we were now living with Carl the Caribou, Ernie the Elk, Gordon the Goat, Wally the Wolverine, and Bob the Bear. The second was that whoever described the genders as being opposite sexes was accurate. We were very different, and this was a good example. I kill it, conquer it, and slap it on the wall as a trophy. She names it and has a relationship with it.

Our first year together was very sweet. God gave us the foundations of love and trust that have lasted many years. We also began praying and working toward an eventual move to Siberia. Two trips fueled our vision that year. The first was to a Global Church Consultation in Seoul, South Korea. It was an incredible time of discovering a bigger picture of what God was doing all around the world. The second was a vision trip to northern Siberia with two friends, John Lamb and Patty. Patty is one of Annette's best friends and would eventually move with us to Siberia.

We started in Anchorage, Alaska, flew to Khabarovsk in the far east of Russia, and from there to Yakutia, the coldest inhabited place on earth. We were still trying to recruit a team, so maybe this wasn't the best place to start. It did give us a better understanding of the people and how we needed to prepare. It also gave us a few opportunities to connect with individuals in unlikely places. We met with a few men in a gulag outside the town of Yakutsk. This was a sad portal into a part of Soviet society that was meant to be hidden. In some ways, it was a microcosm that represented the people of Russia as a whole. It was dark, oppressive, and kept people from experiencing true freedom.

The other encounter for me on that trip came on the flight home. Our Aeroflot flight left Khabarovsk and about ten hours later landed in Anchorage. In 1995, most Americans still enjoyed a platform in Russia simply because of their passports. It was easy to approach people, start conversations, and find folks who were interested in anything we had to say. By and large, those days are now gone, but it was a different time. Also, it was before the September 11 terrorist attacks, so air travel was much more relaxed.

Using the platform of my USA passport and the fact that I had a pilot's license, I asked the flight attendant whether I could visit the cockpit and meet the pilots. Granted, I was a pilot with experience only in single-engine aircraft and only one hundred hours at that, but it was enough to get me an escort up the aisle. She beat on the cockpit door and introduced me (by ethnicity and fellow aviator) to the crew inside. Five friendly faces looked up at me. The flight engineer quickly jumped up and offered me his seat, which was between the captain's and his copilot's seats and slightly back. Behind his seat were two more, one facing each side of the fuselage. One was for the navigator and the other for someone doing radio communication. The captain spoke good English; he instantly removed his headset and struck up a conversation with me. With the autopilot system on, he wasn't distracted, and I seemed to be a welcome break from the monotony of a long transpacific flight.

We told each other about our families and flying experiences. Then he asked the purpose of my visit to Russia. I explained what the trip to Yakutsk was for and that we planned to move to Siberia in a year or so to work as missionaries and tell people about God's love. What he said next instantly made me aware that God had ordained this time in the cockpit. He told me about his wife and daughter in Khabarovsk and said they had recently started attending a new church some missionaries had started. He said he didn't understand what had changed for them and why they wanted him to join them. I knew he was looking for any insight into the changes happening in his family. I told him I thought I might be able to explain what had happened to them because something similar had happened to me a few years prior. I asked whether I could share something from a booklet I had. It was bilingual, so he could read it in Russian as well as in English; it explained how we could have a relationship with God.

At this point, the others wanted to see what I had pulled from my pocket and leaned over my shoulders. The others listened as the captain translated my words, and I explained about God's love and forgiveness. Talk about a great selfie moment, but it was a decade too early. I would love to have a photo of that time, which isn't fading in my mind. What a fantastic place to share with these men, at thirty thousand feet somewhere over the Bering Strait from the flight engineer's seat in an Aeroflot jet! I finally told the captain that my guess was that his wife and daughter had started relationships with God and wanted him to do so as well. I don't know what decisions were made that day, but I have a hunch that the great adventure for one or more of them may have begun right there in that cockpit.

The first year of our marriage had been packed full, and the second year started the same, but it was even better. Annette was pregnant with our first of three. Andrew was born in November 1996. Our Siberian team was growing. Enter the great adventure of parenthood! What a glorious and challenging time. After some cross-cultural training and one semester of Russian language packed into a summer, we moved to Siberia in August 1997. Andrew was just nine months old. He would take his first steps, speak his first words, and hopefully grow some hair on a different continent. Our location was probably harder on grandparents than on anyone else. Annette and I were both blessed to have parents who were so supportive of where we were moving with their grandson.

The move itself was a lot of work. Our dear friend and teammate Patty joined us. The rest of the team was rounded out with three short-term interns from Montana. What a blessing to have them join us for the first year! CRU redirected us, so our new homes would be in a different Siberian city. Six of us, seven with Andrew, moved to Irkutsk instead of Yakutsk. This change meant a bigger city situated on the Trans-Siberian Railroad. Although still very cold, it was warmer than the coldest inhabited place on the planet. If you have ever played the board game Risk, then Irkutsk may sound familiar. The city had nearly seven hundred thousand people, with about one hundred thousand of those being students, who attended about thirty universities and colleges. Students came to Irkutsk to study from all over Siberia, so it was a perfect place to start a student ministry and help plant a church. We were allowed into the country on student visas, which meant spending about nine hours each week at the Pedagogical Institute while improving our Russian. Obtaining visas each year was always a challenge and opportunity to trust God. Patty prayed each year, "God, if you want us to stay here, then it's your job to give us a visa!"

The actual visas were made of cardstock paper with our names, passport numbers, and type of visa printed on them. There was an exit portion of the visa and a reentry portion, which allowed the holder to leave the country once during the year. Though Andrew had his own passport, he didn't have his own visa; instead, he was listed as an infant on Annette's visa.

In January 1998, our team was invited to attend a small conference in Estonia. Tallinn, Estonia, is situated on the Baltic Sea in Eastern Europe. This region was considered more modern and had some luxuries most of Russia and certainly Siberia didn't. This would be a nice retreat and a chance to be refreshed. We had flown to Moscow and took an overnight train to Tallinn. We had a wonderful time with several other teams from across Russia, ate out in some restaurants, and really enjoyed the five days. We boarded the train in Tallinn for our return to Moscow, loaded down with supplies unavailable in Russia. We managed to get all our luggage, plus a baby stroller, into our cabin on the train. We weren't traveling light. Annette and I tried to sleep as the train car rocked Andrew into a slumber.

At about two a.m., we made it to the border, and Russian customs and passport control woke us. I handed them all three passports and both of our visas, hoping for a quick stamp and quiet to avoid waking Andrew. After a few minutes, the woman in uniform asked for something else. I didn't understand, but she motioned to Andrew and said, "Visa." I pointed to Annette's visa. His name and passport number had been on the exit portion of her visa but had never been printed on the reentry portion. Through tears, Annette pleaded with her to allow us to continue and get it straightened out in Moscow. Annette lost her case, and no amount of tears or a sleeping baby or pleading from the ignorant foreigners was going to change this woman's mind. This wasn't the last time I was tempted to offer a bribe to an official in Russia. Later, I stopped using the term *bribe* and instead considered it a "gift given in secret."

Off the train, we went. About then I remember thinking, *I have had about enough adventure!*

Exhausted and with a cranky baby and a wife in tears, I chucked our overstuffed suitcases and bags off the train and into a blizzard. There was a cold warehouse not far away, so the first order of business was to get my wife and child some shelter. Eventually, we had all our belongings in the freezing concrete building and waited for a train going back to Tallinn. About four hours later, we were able to muster the energy to load it all back on a westbound train. When I think of that night, "miserable nightmare" still comes to mind.

Back we went to the hotel and McDonald's in Tallinn. Our son might have been only fifteen months old, but he was well acquainted with Happy Meals. That trip to Estonia sealed the deal with Andrew and McDonald's—so much so that for years later, while in Irkutsk, he asked me whether we could get on an airplane and fly to Moscow so he could go to McDonald's. The closest one was five time zones away. Now that is good McDonald's marketing!

*Andrew and Annette celebrating a visa with a Happy Meal.*

After a few days and many calls with our leadership in Moscow, I was at the Russian embassy in Tallinn, securing a visa for Andrew. Someone had to pull some strings somewhere for all this to happen. Annette and I were still on student visas, while Andrew reentered Russia a few days later with a business visa. My fifteen-month-old son was a businessman in the eyes of the Kremlin. I was beginning to see there was always an alternative way to getting things done in Russia.

CHAPTER 13

# Always Another Way

As you can tell by reading the last twelve chapters, I wouldn't consider myself to be a strict rule follower. In fact, I would subscribe to the theory that "it's easier to ask forgiveness than permission." But I still come from a culture where, generally speaking, there is a correct way to do things, and when it comes to legalities, there is only one way. For anyone who has traveled or lived abroad much, you know this isn't true in many countries. Most Russians operate under the philosophy of "That's one way to get it done." In other words, there is always another way, and if you scratch my back, I will scratch yours. In Russia, a lot of scratching is done with vodka or cash.

Of course, I was under the impression that bribes were illegal and just wrong. But I read in Proverbs 21:14 (NIV), "A gift given in secret soothes anger, and a bribe concealed in the cloak pacifies great wrath." So there is another way to look at it, especially in a culture that has worked this way for centuries. "When in Rome …" Maybe I needed to enculturate and embrace instead of judging if I wanted to get anything done. It's hard to change your thinking about things like this quickly, so I usually wasn't looking for "the other way" in most situations. But occasionally, it was quite clear to me that there was another option. For instance, when I was pulled over for traffic violations, the police usually issued a ticket, and the fine was payable immediately. The price for most offenses was around fifty rubles or about seven dollars (not a huge deterrent). Unless you didn't need a receipt for the payment and didn't want a copy of the ticket, the price went down to thirty rubles. You get the picture. There is always another option in Russia, and I made sure I didn't get behind the wheel without thirty rubles in my pocket.

One of my first chores in Irkutsk was to purchase a car. The handful of other missionaries we knew didn't own vehicles and simply used taxis or public transportation. But after my experience in Mongolia, I knew that having our own car would make life a lot easier. Shopping, going to the university, hauling students around, and so forth—all of it was easier with a car.

The problem wasn't purchasing a car but getting a driver's license. At that time, Russia didn't accept foreign driver's licenses or even the AAA International License like they do today. So if you wanted to drive legally, you needed a Russian license. But to obtain one, you needed to first pass the written exam and a driver's test with the examiner. The test had to be taken in Russian, and no translators or translation of the exam was permissible. You start to see the problem. Unless you read, wrote, and spoke impeccable Russian, you would never pass these tests. In fact, many Russians couldn't pass the exam simply because there were so many traffic laws, and many people didn't know them all. It became clear that the DMV didn't really want anyone driving, and there must have been miracle after miracle that allowed for so many Russian drivers to actually have licenses ... unless there was another way.

When I asked my friend Oleg, he said, "There must be another way to get a driver's license." He inquired at the office of motor vehicles and was told there was indeed another way. First, I needed an official translation of my Colorado driver's license, then have that translation notarized and bring that with some cash to the office. I think it was $500 US. Whoa, that is pricey. But the alternative was to wait for years and hope I could pass the tests. I took $500 out of the car budget and went to get my license translated and notarized. I was relieved to have another way to get a license and soon begin searching for a vehicle.

A week later, Oleg and I walked into a room about half full of people waiting to take their written exams. Oleg handed over the translation and copy of my Colorado license, then waited. The examiner nodded, which was Oleg's cue to hand him the cash. It was done on the sly, and I began wondering how legit this other way was. On the one hand, $500 given in a handshake wasn't what I was expecting, but if all they wanted was cash, then why require a translation of my license?

Oleg returned to me and said we needed to sit down at the long table with the others. By now I was used to not always knowing what was going on and living with some ambiguity. *They must be printing my new license*, I thought. *I'm sure that's the holdup.* But that thought was dashed when the examiner entered the room again with a stack of exams. He handed one to each of us who were sitting at the table, except Oleg.

I was now a bit concerned that I had just been taken for $500. Oleg and I looked at each other, and his shrug told me he also was confused; this wasn't playing out like he had expected. We were then instructed to write our names on the first page of the exam, which had three pages of multiple-choice questions. I did as instructed and then raised my eyebrows to Oleg. Sitting next to him, I showed him the exam. He read the first question, and I could see, despite him having his own license for years, he hadn't a clue as to the correct answer. The examiner was now circling the big table with a watchful eye on us all. As he looped around behind me, he whispered "D, D, C, A." Did that just happen? I tried hard not to show my surprise or to write four answers so quickly that it would tip off the other test takers that I hadn't even read the questions.

I slowly shot a look and a grin at Oleg, then wrote down the answers to the first four questions. Four at a time, the examiner fed us the answers until I was the first one finished. In a weird way, I felt proud; I had never been the first one to finish a test before. Eventually, everyone finished and handed in his or her exams, and then we waited. About fifteen minutes later, the examiner appeared in the room again. The anticipation was shared by all to see whether we passed. He approached me first and loudly proclaimed, "Congratulations, you got a hundred percent!" Everyone in that room knew what had just happened, but we all played along like we had read from the same script. A few minutes later, I was posing for the photo of my very own Russian driver's license. There were indeed two ways to get one, and mine came through door number two.

Living in the Russian culture proved to be a great adventure in and of itself. There were so many unknowns, and with each one came an opportunity to trust God. One challenge for many missionaries is finding hobbies they enjoy. We were told in our cross-cultural training that finding things we enjoyed doing would be a key to our longevity. Of course, my hobbies revolved around the outdoors and required a lot to participate in them. To go hunting in Siberia required a lot of planning, some purchases, and not a little patience to make them happen. I had found the equivalent of our fish and game office not long after moving to Irkutsk.

Step one was figuring out seasons and game laws and, if possible, meeting some other hunters. They say a picture is worth a thousand words. So, during my first visit to the game and fish office, I took a photo album of hunting pictures with me. I'm pretty sure I didn't know a thousand words at that point, but the photos spoke what I could not. Hunters are similar in just about every culture. We all love to flash a photo, tell a story, or, when possible, present the kill. The game wardens in this office were no different, and I achieved near-celebrity status after that first visit.

One older gentleman, in particular, showed a lot of interest in getting to know me further. Alexander was grandfatherly and spoke a little English. He loved to practice the little he knew and was patient with me while I spoke Russian. Over the next eight years, we spent many hours in his office and talked about life, hunting, God, and his new wife. I had missed hunting season that first year, so I would need to wait nearly a year before being given the chance to pursue my hobby.

I also needed a gun if I was going to hunt. But as a foreigner, I wasn't allowed to purchase or possess one. Door number two opened again with Oleg. Since he wanted to learn how to hunt, he suggested I pay for a hunting rifle, and he would put it in his name and keep it at his house. Since I trusted him, this seemed like a good way to take steps toward my favorite hobby and spend time with a friend. Soon we purchased a rifle, scope, and several boxes of ammo, and I found that Oleg loved shooting *his* new gun. We had planned and purchased, and now it was time for patience. We had to wait many months before we could actually hunt. As the months went by, I had a chance to learn more about the Russian game and fish from Alexander. I was amazed to find out there were as many game laws as there were traffic laws. Few people, if any, actually obeyed these laws, but they existed. As a favor to Alexander, I even edited the English version of some of their big-game laws for hunters who might come from abroad.

Since I was familiar with elk hunting and there were elk not too far from the city, this was my first choice the next fall. Like in Colorado, the elk in the Irkutsk region begin their breeding season around the middle of September. By the first of October, they were at their peak of the mating season, which made it easier to locate bulls since they were very vocal.

I strolled into Alexander's office on the first of October, not having seen him in several months. We greeted and sat in his office to chat. He was the deputy director for the entire province and was well known and had friends in high places, but he never once acted like it. He had a genuine care and concern for our family and even considered himself like a Russian grandpa to our kids. After catching up, I asked him about getting a hunting license like we had discussed the year before. I had a long weekend coming up and wanted to take advantage of it and spend some time with Oleg in the woods, chasing elk. After not hunting for two years, I was pining.

Alexander's smile disappeared and with it my enthusiasm. I knew something was wrong, and he finally explained that the hunting season for elk had already ended. He said they had closed it earlier that year than in previous years. I was crushed. The only things I could hunt were some birds or waterfowl, and I didn't have the right gun for those. The thought of waiting another full year was a hard pill to swallow. I sank in my chair, my disappointment obvious. Alexander paused for a moment in thought and then said, "But let me see what I can do."

*Let you see what you can do?* I thought. *What does that even mean?* He tried to reassure me and said, "Just come back tomorrow." I left his office without much hope, thinking I needed to find another hobby.

I did as I was told, and the next afternoon, I was back in his office. He excused himself for a minute and went down the hall to the director's office. I could hear him ask for a paper. I heard the unmistakable sound of an official stamp being placed on a document. Alexander reappeared with a full one-page document, signed by him and his boss, and stamped with the official emblem of the game and fish office. He handed it to me and explained that this was like a license—emphasis on *like*. I looked at the official letterhead, stamp, and signatures. As I scanned down the page, I saw my name in bold letters and the license plate number of my car. The rest of the document was written in formal Russian, and with my limited language ability, I couldn't read it.

I glanced up at Alexander again with an inquiring look. This piece of paper didn't resemble the other hunting licenses I had seen. He read my mind and answered my question with the same statement, that this was *like* a license. I decided I needed further clarification, so I asked if I could shoot an elk with this. "Yes!"

"Can I shoot a moose?"

"Yes."

"Can I shoot a bear?"

"Yes, yes, yes. You can shoot whatever you want!" His reply was direct, intentionally combatting my disbelief.

Wow! This was like winning the hunting lottery. I was suddenly back in the game and raring to go. I thanked him immensely and left his office, calling Oleg as I walked to the car. He would be ready the next day, so it was time to pack supplies and food while Oleg brought the gun and ammo. Before hanging up, Oleg asked whether I had gotten a hunting license.

I said, "Yes, sort of. Alexander said it's *like* a license." Twelve hours later, filled with enthusiasm, we were on our way out of the city. Hopes ran high for Oleg since this was his first hunt ever, and I was overjoyed to get back to doing something near and dear to my heart.

I pulled out the map as I drove and handed it to Oleg, pointing to the spot Alexander suggested we go. He studied it for a while, then asked whether he could see the hunting license. Sounding like I was mimicking Alexander, I explained again that it was *like* a license as I pulled it from the console between the seats. He removed it from the protective sleeve and began reading.

After about thirty seconds, he began chuckling and then affirmed what I had been telling him that this was in fact NOT a license. I agreed and said again, "Alexander said it's *like* a license."

Oleg turned toward me in the car and with a wry smile divulged what had been hidden from me. "Spencer, he made us deputy game wardens for the weekend!"

I'm not sure why I was surprised; I thought I had learned that there were always two ways to get something done in Russia. The irony was the literal translation of "game warden" in Russian means "protector of wildlife." I had to admire Alexander's creativity. The other ironic part of this was that a few years earlier, I had wanted to become a game warden. I was one now, at least for the weekend.

I continued to be able to hunt while I lived in Irkutsk and enjoyed the refreshment from time to time in the woods. Nearly every year after this, I had a real hunting license. Alexander made sure. On one of my visits to his office, I took with me some of my elk calls. Elk are very vocal animals, and both the bulls and cows "talk" a lot. Once you learn their language a bit, it can be very helpful when hunting.

Alexander gathered some of the game wardens into his office one day as I gave a demonstration with the calls. They all loved it and immediately recognized the huge advantage it would be to call like I was. The problem for them was that they had no access to calls like I had, nowhere to buy them. I knew then how I could show my gratitude to Alexander. During the next trip to the United States, I would purchase a handful of calls for him. He could use them or give them as gifts to others. So, a few months later, I bought a half dozen cow elk calls and bugles in a sporting goods store.

A month later, I headed for the fish and game office and handed Alexander a bag full of elk calls. He reacted as if I had handed him a bag of cash. He loved my gift! I told him to use one and give the rest to whomever he wanted. It felt good to be able to give something to him after all he had done for me. I visited him over the next few months but didn't hear what he had done with the elk calls. Then, during one visit to his office, he shocked me with some news.

He told me the KGB had come to see him the week before. Even though the KGB didn't officially exist any longer, many older Russians still referred to them as such. The new name was FSB (Federal Security Service). Since Alexander loved a good joke, I waited for the punch line. It didn't come, so I finally asked what the KGB had wanted. With great animation, he said these three agents strolled into his office and simply asked, "Where is your Spencer?"

He now had my attention. As a foreigner, nothing good could come from the FSB wanting me. I was instantly sweating. Being very protective of me and our friendship, Alexander asked why they wanted to know where I was. Sheepishly, one of them spoke up and replied, "We hear he has elk calls." Are you kidding me? Ex-KGB officers really want elk calls? Knowing he had something they wanted, Alexander said, "Yes, he can get elk calls, but it will cost you."

I couldn't believe my ears. I wasn't in the import business and had no intention of making a buck off the FSB. I then inquired how they had heard about me or the calls I had. I followed it up by asking him to whom he had given the calls I gave him months earlier. He looked toward the ceiling as if he were trying to remember and at the same time not wanting to give away just how many friends in high places he had. One call he had given to the owner and CEO of the vodka plant in town. Hmm, that explained one.

Who else was blowing these calls I had brought fifteen time zones? He then revealed that maybe he had given one to the chief of police.

That explained more. No wonder the word was out. Not wanting to wait a full year until my next trip to the States to satisfy the growing demand I had created, I asked my brother to send me a handful by post. About three weeks later, I delivered the elk call contraband to Alexander with strict instructions to *give* them, not *sell* them, to the KGB. Ultimately, I knew God was our protector and always watching over us, but if I had the chance to scratch the backs of some FSB/KGB officers with an elk call, I would do so.

CHAPTER 14

# Joys and Pain

During our time in Irkutsk, eight years in all, we always felt cared for. It became clear to us as the years went by that God was indeed watching over us—physically, spiritually, and emotionally. That isn't to say we didn't have tough times or days when we wanted to pack it in and leave. Annette and I called those our "I hate Russia days." Fortunately, we seldom had those feelings on the same day, so we could pick each other up. Some of the troubles other foreigners experienced just never seemed to come to our door.

Russians use the word *kreesha*, which is the word for "roof," but the slang meaning is for someone who offers you protection. In a culture where power is everything and abused often, having a *kreesha* is highly beneficial. Some people pay to have this protection, and it is extorted from others. At any rate, our Russian friends would often ask who our *kreesha* was, because they noticed we were never harassed or bothered. We usually told them God was our *kreesha*. But at times, God uses people to carry out his bidding. Looking back on this now, I see that God may have used the ex-KGB guys with elk calls to make sure we were seldom bothered, broken into, and robbed only once. I may never know for certain through whom his protection came, but I do know that ultimately we trusted God.

Part of the adventure of walking with God is that so much is unseen. Adventure cannot be completely planned, even though we try. This fact was driven home time and again during those eight years. There was so much going on behind the scenes that we were never aware of. The spiritual world is mostly unseen. This is where faith is exercised; the trust God wants from us comes by not always seeing what is next. If we could see very far, we might not continue. Faith is what pleases God and where we learn to rest. A mixture of good and bad, of pains and joys, is the field on which faith is practiced. Would I lean into my heavenly Father during both the ups and downs? Russia was the field where some of the greatest joys and deepest pains came about for Annette and me. It's easy to call the joys part of the adventure, but it's a bit more difficult to include the pains. Both passed through his hands before coming into our lives.

Our family had settled into a rental apartment in Irkutsk and made it home. We lived on the sixth floor of a concrete building that looked identical to one thousand other buildings in the city. It was here where Andrew learned to walk and start talking in both English and Russian. It was also the place where we experienced a great deal of pain, the kind that would shake me and leave me on my knees.

We had talked and prayed about having more children, so after our first year in-country, we decided it was time to try. Annette became pregnant with our second. Our plan was to wait until the eight-month mark and then go back to the States on furlough for the birth. What we had seen or heard concerning the hospitals in Irkutsk wasn't good. Even our Russian friends cautioned us not to go to a hospital unless we wanted to catch something. Our faith was fairly strong but not strong enough to have Annette give birth there. And given my experience a few years earlier in a Mongolian hospital, we figured, *Why chance it?*

There were only a handful of missionaries in Irkutsk in the '90s, and we all seemed to know one another and enjoy fellowship. One of our friends and fellow missionaries was a medical doctor, who had a family practice before moving to the mission field. She had graciously agreed to give Annette some prenatal care. We were so thankful to have her in the city just a phone call away. The first trimester for

Annette came and went with only a little nausea. Annette's exams were all thumbs-up. We were into the second trimester when Annette had another exam. Annette had experienced some changes, and this time the doctor came back with the devastating news that Annette was going to miscarry; it was inevitable.

We were both shocked, scared, and deeply saddened. We were aware that miscarriages weren't uncommon, but that fact was of little comfort at the time. The doctor gave us instructions to stay at home and be ready because when it actually happened, we didn't want Annette anywhere but at home. About midnight, it began. Thanks to her coaching, we had some idea of what to expect. My job was to make sure she was comfortable and keep an eye on her.

Before leaving the United States, one of the preparations we felt necessary was for someone on our team to have at least some first-responder training. I had been able to attend wilderness EMT training the summer before moving to Siberia. The Wilderness Medicine Institute had also donated a medical kit stocked like an ambulance. During that first year, I didn't have much need to use my training. That ended on this particular night in our apartment. I now felt ill prepared with my wife being my patient. Nearly an hour after our baby appeared, it became clear that something was wrong. The bleeding didn't slow down.

I was now checking her vitals every fifteen minutes. I was also on the phone with a doctor in the United States, who patiently stayed on the line with me for about an hour. I then decided to make the two a.m. wake-up call to our doctor friend in the city. She was at our flat in about thirty minutes. Annette's bleeding continued, and her blood pressure started to drop. She was faint at times and eventually even passed out.

I saw the panic on our friend's face, and after a quick exam, she ordered me to get my car ready. We needed to get her to the hospital because there were complications, and her bleeding wouldn't stop. She would need to have surgery in order to stop the bleeding and save her life. A dilation and curettage is a procedure to clean tissue from the uterus. I called our teammate Patty to come and stay with Andrew and then rushed to get our car in front of the building. We placed a chair in the elevator for Annette to sit on because she could no longer stand, and I couldn't hold her that long. Once in the car, with the gals in the back seat, I sped toward the Russian hospital to face other fears.

Sometimes fears get blown out of proportion in our minds. When we face them, we realize they aren't as bad as we anticipated. And while the hospital was certainly not the Mayo Clinic, it also wasn't like the hospitals I'd had the privilege of visiting in Mongolia. The regional hospital was a huge building, about ten stories tall, with several wings stretching out from the central original building. Upon our arrival, it was difficult to find an entrance that was unlocked and hopefully led to the ER. Finally, we found a door unlocked and a nurse nearby who found us a rickety, old wheelchair. We scooped up Annette and started trying to communicate her condition to the nurse.

When you study another language, it usually begins with conversational vocabulary. Words like *miscarriage, uterus,* and *fainting* usually aren't encountered in conversations. I think both our doctor friend and I probably had Oscar-worthy performances, at least in the charades category. Finding a gurney to put her on, the nurse quickly pointed us to the elevator and said, "Go to the sixth floor." The doors opened, and a scarf-covered babushka waved us inside. She pushed the button, and up we went. Since she wasn't wearing any nurse's attire or uniform, I wasn't sure whether she was just a helpful soul or this was her job. But her face showed concern for us and a warmth that was hard to see on most faces in Russia. The doors opened, and we wheeled Annette down a hallway in the direction the old woman pointed. The hall was empty, and there were no wall hangings or decorations, no signs to identify that we were even where we needed to be. Just old and yellowed-looking tiles; dim, flickering lights; and a couple of chairs along one wall greeted us. We raced down the hall, one of us pulling the gurney and the other pushing. We nearly passed the open door on our right with a doctor standing a few feet back. As we backed up and pulled Annette inside, it became clear that we were in an operating room. Later I learned it was an OR on the gynecological floor.

I stepped back from Annette's side as the Russian doctor and our friend began talking while they started an exam. I suddenly felt useless. All I could do was pray. I was motioned to go back into the hallway, and someone closed the door behind me. In a few seconds, our friend poked her head out and said they would allow her to stay with Annette during the operation. I was thankful for that. She then instructed me to go back home and get some sheets, a blanket, and a pillow for Annette. I was dazed, and this information just sort of bounced off me. She repeated herself and then added, "Trust me. You do not want your wife lying on these mattresses without a sheet!" I then recalled what some of our Russian friends had said about not going to the hospital unless you want to catch something. Convinced she had made her point, her head disappeared back into the OR.

I have since learned much more about miscarriages. But at that point, I understood very little. I knew only that the day before, we had been pregnant with our second child, and now we weren't. I knew that, at least early in a pregnancy, it wasn't uncommon to miscarry. I even knew that some women miscarry without ever knowing they were pregnant. What I didn't know was just how developed a baby was in the second trimester. Oh, I had seen pictures in textbooks, but that didn't come close to preparing me for seeing and holding my son. We would later name him Jack. That was over twenty-three years ago now, and the tears still flow freely.

Before leaving our apartment that night, our friend suggested taking the baby with us, since the doctors may be able to examine him and discover what caused the miscarriage. For an instant, I froze when she told me to do this. I didn't know whether I even wanted to attempt this or how to. What did I use? What did I put my baby in to take him with us? I went to the kitchen and grabbed the first container I found. After all, this wasn't the highest priority now; I just wanted to get help for Annette. But as we left that night, I managed to tuck the container with our lifeless baby under my arm.

Now I was sitting in that dimly lit hallway on a wooden bench, with my wife in the room across from me and our baby in a container inside a plastic bag on the bench next to me. As I tried to wrap my head around where I was and where Annette was, I could no longer hold it together. I wept. I crumbled. To this day, I have never felt more alone in my life. There was a numbness mixed with my pain. I cried out to the Lord again. "God, will you spare her life?" I was hopeful, but my EMT training also brought the reality of the situation into focus. I knew she was near the passing point.

Something jolted me out of my deep thoughts and back to the task at hand. I needed to drive home and get the clean bedding before she came out of surgery. I forced myself off the bench and down the hall, which brought the painful distance from Annette. I couldn't leave, but I had to. I pushed the elevator button and waited. The doors opened, and that warm and wrinkled face was waiting for me. I stepped in and toward the babushka as she slowly pushed the button to take me down.

In 1998, there were seven hundred thousand people who called Irkutsk home. At that time there were about five or six churches I was aware of, and most of those were small and recently planted. I estimated there were no more than five to six hundred people attending those churches on a given Sunday. That is less than .001 percent; that's not very good odds of randomly running into someone in Irkutsk who attended a church. With more bedside manner than I have ever experienced, the old woman asked me whether the woman we had brought in was going to be okay. It would have been hard for me to speak in English at that moment, so mustering my Russian was nearly impossible. I managed to say I was praying for God to take care of her and that she was my wife.

She then asked whether I was a believer. I had been asked this a couple of times during those first few years in-country, and it was usually whispered in my ear. During the communist era, many Christians were persecuted for their faith, some were killed, some were imprisoned, and many were shipped to Siberia. Even though it was eight years after the collapse of the Soviet Union and there was a perceived religious freedom, an inquiry about your faith was still done quietly and discreetly.

Since we were alone on an elevator, her question wasn't a full whisper in my ear. I immediately replied, "Yes, I am a believer," and I asked whether she was one too. She wanted further clarification, so she asked whether I was a believer in Christ. I confirmed that I was and was also a missionary here in Irkutsk. I asked her whether she was also a follower of Christ, and she smiled and nodded. It was now about three o'clock on Sunday morning. I asked her whether she attended a church, and if so, which one?

She said, "I go to the church over on Kaiskaya Street." I was familiar with the church and had visited it numerous times. She then added that she would be going to the service in a few hours when her shift ended at eight a.m. I asked whether she would pray for Annette and whether she would ask the church to pray for her as well. She reached for my hand and said, "Of course."

As I walked to my car, I sensed a comfort and closeness to God I have rarely felt. It was as if God himself had wrapped his arms around me and pulled me close. I knew he was there; he was aware and fully engaged. My despair and panic began to subside, and I calmly drove home and retrieved the bedding. I arrived back on the sixth floor just in time too. A nurse was pushing a bed on wheels into the OR. The three-inch-thick mattress was worse than I could have imagined. It needed to be burned. She helped me quickly put the sheets on it as I tried hard not to stare at the old bloodstains defiling it. I had no idea that hospitals didn't provide bedding and was so thankful my wife wasn't lying directly on that mattress.

When I saw Annette, she was still under anesthesia and not fully awake. But our dear doctor friend said the surgery went well and that she would make a full recovery. I knew she was exhausted, having been up all night. We both were. There were other things I soon found out; the hospital didn't provide their patients with food and drinks. Back home I went to get soup, crackers, and drinks. The round trip took about forty-five minutes, and I was back and headed for the sixth floor again.

Other than when her dad walked her down the aisle at our wedding, I don't think I have ever been more excited to see my wife. It took a little longer to find her when I returned because they had moved her to a recovery room with six other women. When I saw her, I knew we were going to make it through this. God's grace would see us the rest of the way.

She started getting stronger over the next couple of days and really wanted to go home. I arrived on the second day and entered the room again. All the same faces smiled at me as I walked in, except Annette's. I asked whether anyone knew where she was. They all shrugged or shook their heads. Uh-oh! My wife might have gone AWOL, and I wouldn't have blamed her considering the conditions.

Back into the hallway, I went and headed toward the restrooms; maybe she was there. Suddenly, from behind a long curtain, I heard a whisper. I looked and saw her peering around the edge, only half of her face visible.

"What are you doing, and why aren't you in bed?" I asked.

She said she was hiding from the "shot lady."

"The who?"

"Shot lady!"

Evidently, a woman dropped by their room twice a day with a tray full of syringes and gave each of the ladies about four different injections. I asked what they were. Annette had no idea. We later found out another major difference in Western and Russian medicine. Everyone in the hospital got large doses of antibiotics and vitamins, no matter what. Because sterilization and hygiene aren't as valued, this is the way they deal with the consequences. Annette was able to go home the next day, miraculously without any sicknesses. Praise the Lord!

Some wounds aren't meant to heal, at least not while we are on this earth. Losing Jack and nearly losing Annette will be two of those wounds. Even though the Lord brings healing and badly needed perspective, those events still hurt. The pain was eased more when Annette became pregnant with Kate about a year later. We both had worked through the pain, albeit quite differently. I desperately wanted to hold another baby. I couldn't handle thinking that the last baby I would hold was Jack. And so God blessed us with Kate. But the journey with her in the womb proved to be quite difficult as well. It was a year later, and Annette was in her second trimester again when she began to experience symptoms that were similar to the miscarriage the year before.

SOS insurance had just been a line item on our budget, for which we had raised financial support. It wasn't a big expenditure but one we were aware of; we hoped we'd never find out how good the insurance actually was. When someone says, "SOS" I immediately think of someone stranded on an island and SOS being spelled out on a beach with logs. Or I think of lyrics to a Police song. This SOS insurance guaranteed medical evacuation from anywhere in the world; within twenty-four hours, it would have us at a medical facility with Western- or American-trained personnel.

Since moving to Irkutsk, I had learned a bit about how the Russian government felt about their airspace. As you might imagine, they were very protective. The result of this was twofold. One, it meant they were never going to allow a foreigner like me to fly a US-made plane in their airspace. Dashed were the hopes of using my pilot skills to spread the message in Siberia. Two, it made for a two-hour-longer medical evacuation. Because we lived on the continent of Asia, Seoul, South Korea was much closer to us than Europe. But the only medical jet and personnel allowed to fly in Russian airspace were based in Moscow. That meant a six-hour (five time zones) flight just to get to Irkutsk. Then it meant another seven-hour (six time zones) flight to the nearest hospital in Europe.

Our organization's leadership didn't hesitate to get SOS involved, and the medevac Jet was dispatched. It brought some relief to us but also many questions. Once again, we found ourselves in a position of trusting God. Of course, we have opportunities to trust God in our daily lives. And at times, I have trusted him, and at times I have chosen not to. But in this situation, it felt like there was nowhere else to turn, nothing else we could do. For about eight hours, we waited and prayed. The evacuation was imminent. A flight for life jet landed with a complete medical team of nurses and a doctor.

Our experience at the Irkutsk hospital the year before had put some of our fears to rest. But since Annette was still pregnant, we wanted to do all we could to get her and the baby to the best care available. Those eight hours were tough; my routine was, pray and then pack and play with Andrew. Repeat again in a few minutes. I realized then that the wounds of a year ago hadn't healed and probably never would. This time I found myself praying for God to spare not only Annette's life again but also our baby's.

Packing for this trip was like no packing I had ever done. Normally, Annette would never let me pack for her. It was more like turning dresser drawers upside down in a suitcase than actual packing. Annette was on bed rest, so the fashion-challenged, color-blind husband was on duty. That was the first packing hurdle we had to overcome. The others were that I didn't know where I was packing for her to go, how long she would be there, or whether Andrew and I would be allowed to accompany her. So many questions ended in prayer. In the end, I packed a bag for each of us and prayed there was room on this small jet for more than just Annette. A teammate drove us to the airport and out on the tarmac right up to the jet. It was then that we learned the destination of this flight—Helsinki, Finland—and that there was room for Andrew and me too.

Annette was made comfortable, and the staff continually monitored her on the flight. We landed in Helsinki about eight hours later after a stop to refuel and do passport control in a tiny village near the Arctic Circle. A flight to Seoul, Hong Kong, or even Bangkok surely would have been shorter. It was mid-December, however, and Helsinki was beautifully decorated. The ambulance whisked Annette and us to the hospital. The hospital staff in Helsinki was wonderful. We couldn't have asked for better care.

Our fears were relieved after the initial examination, when they found both Annette and the baby would be fine. After a couple of days in the hospital and then another few days of rest in a hotel, we would be allowed to fly back to the United States. It was also decided then that Annette should be in the States close to her doctor for the remainder of the pregnancy. It turned out I didn't pack enough for any of us. One small bag of clothes wasn't going to cut it for six months. Her due date was in June, and the baby would need to be two months old before we could return to Russia. Some new challenges lay ahead of us, but at least Annette and the baby were out of danger.

On May 15, four weeks early and only weighing five pounds, Katie arrived via emergency C-section. The girl just couldn't wait any longer. After seeing the danger both she and Annette were in for this delivery, I could only fall on my knees and thank God that he had brought us back to Colorado and a good hospital and great doctors for this event. I couldn't imagine being in Irkutsk for Katie's delivery. While there was still pain felt with the loss of Jack, some of it was washed away with the birth of Katie. In Isaiah 61:1 (NLT), we are told that one of the reasons for Christ's coming is to "comfort the brokenhearted." Not only does God see and know about our heart-felt pains, but he wants to bring healing as well. Even though there is still pain, it was mixed with healing and now the joy of having a daughter. I was learning that part of this adventure with God meant letting him be my comfort and peace while joy and pain coexisted. God had blessed us with the incredible joy and pain of three children.

Chapter 15

# Planes, Trains, and Taxis

Faith is exercised every time we step on an airplane. It's true for everyone, even though we sometimes don't even think about it. Air travel had become as common to most of us as the horse to most Mongolians. And even after flying over a hundred times all over the world, I still find myself praying each time I'm on a plane that takes off and lands. It's second nature for me now. This doesn't rise out of fear as much as from faith combined with the logic of gravity. Being a pilot doesn't really make it easier because I know the amount of human error that can exist on any flight.

Bottom line? I still fly. But this decision is based partly on the fact that I have been to the edge of safe aviation and back now. It's kind of like riding an old, tired mule, which is completely unexcitable after you've been bucked off an ornery, frisky horse. It just feels safe in comparison. Flying on Russian Aeroflot planes in the 1990s was a bit like breaking that wild stallion.

During the Soviet era, Aeroflot was the largest airline in the world. It had a bigger fleet of airplanes and flew more miles than any other aviation company in the world. The state-run Aeroflot began to disband during the collapse of the Soviet Union. Aeroflot still exists but is a fraction of the size it once was. Many of the towns and cities where Aeroflot flew decided to change the aviation industry by seizing control of aircrafts on their runways. It was a free-for-all in many ways and stripped Aeroflot of many of their aircraft, but it gave rise to what many began calling "Baby-Flots." By painting over the word Aeroflot on the tail and fuselage, many of these cities started their own aviation companies with new names. So names like Irkutsk Air, Novosibirsk Air, and Omsk Air all entered the market, thus giving birth to "Baby-Flots." The economy of Russia and former Soviet Republics was in shambles during this time and eventually most businesses were as well. The aviation industry mainly relied on parts robbed from other aircraft because nothing new was being produced. The end of every runway looked like an airplane junkyard. Rows of Tupolev planes were parted out and left like skeletons on tarmacs and runways across the former Soviet Union. Not only did it make for ugly takeoffs and landings, but it also left you wondering what used parts had found their way onto the plane you were on.

Chelyabinskaya Avia, a relatively new "Baby-Flot," was now carrying passengers from Irkutsk to Moscow and back. The town of Chelyabinsk is in the southern part of the Ural Mountains, and in 1998, I had never heard of it. Since we booked tickets to Moscow on this airline, I decided to find it on the map. It was a city that was roughly the size of Irkutsk, so it had also become home to one of the new aviation companies emerging on the scene. We chose to fly on this airline because the flights weren't departing in the middle of the night as most airlines were. We figured it was easier to entertain a well-rested toddler than to try to get one back to sleep. The first thing I noticed as we walked across the tarmac was that the new paint job was already fading and revealing that this plane had also once belonged to Aeroflot.

Hopefully, cosmetics were the extent of the problems on this plane. We found our seats about midway down the aisle. I had the window seat, and Annette sat next to me, allowing us to take turns holding Andrew, who was about two years old. At first, the row of skeleton planes distracted me for a spell. Then my gaze turned down to the tires that were visible and directly under me. I instantly had a flashback to a Mongolian airplane with bald tires and cords starting to fray. The only difference between those tires and these was that these were bigger and had a lot more weight on them. Time to pray and turn my focus elsewhere. We taxied to the end of the not-so-smooth runway, where the rusted hulls of more gutted aircraft came into view. As we picked up speed with the thrust of the engines, I tried not to think about the bologna skins under me. The faster we went, the rougher it got. The plane seemed to shake and rattle as we rolled. Then, as we neared takeoff speed, the rattling increased and seemed deafening. The nose began to lift off as we reached a crescendo of noise.

I was anticipating seeing pieces of rubber being flung from the centrifugal force when I heard a loud crash from inside the fuselage. A five-foot ceiling panel suddenly dropped into the aisle, narrowly missing hands and feet. After clutching Andrew and Annette, I made a vow to never fly Chelyabinskaya Avia again. Once airborne, a flight attendant came back and looked at the panel, at the insulation and now-visible wiring, and simply shrugged. This wasn't her first rodeo, and her edge of safe aviation was obviously much further than mine. Someone finally removed the panel from the aisle, and to this day, I still wonder where they stowed it.

I'm not sure whether Annette slept during that six-hour flight, but I know I didn't. Andrew, oblivious to the dangers at hand, napped the last hour or two. My thoughts and prayers now focused on the runway in Moscow and hoped it was smoother than in Irkutsk. I just didn't think this old bird could handle another bumpy one. It was during the initial descent that we knew something else was wrong. Everyone experiences ear-popping as pressure changes with elevation on a flight, but this was different. I couldn't pop my ears fast enough. As soon as I plugged my nose and tried, I needed to do so again. Then, as if someone had pinched Andrew and woken him from a deep sleep, he started screaming. He was in serious pain, and there was little you can do to help a toddler or baby in those situations. We gave him a pacifier, hoping it might allow his ears to pop.

I knew then that there was a leak and that we were losing cabin pressure. I thought the oxygen masks might drop any second, but they never did. An empty two-liter plastic Coke bottle rolled up next to my feet from somewhere. It took only a few seconds to be crushed like something had sucked all the air out of it. My prayer now was, "Lord, protect my son's ears." After touching down on a relatively smooth runway, the plane slowed, and my heart rate finally did too. Annette and I looked at each other and knew it was time to adopt a new "Baby-Flot."

Train travel was rather mundane compared to air travel in Russia. The Trans-Siberian Railroad ran a few blocks from our apartment building in Irkutsk. It started in Vladivostok in the far east and ran all the way to Moscow and St. Petersburg and on into Europe. This was a much less adventurous method of crossing the country but was much safer too. A two-day ride from Irkutsk to Novosibirsk cured me from ever wanting to take a longer trip. We piled CRU staff and students on the train to go to a conference. The novelty of train travel on the Trans-Siberian wore off quickly, but the time it afforded with all the students was priceless. It was rather peaceful to watch the country roll by, but two days of peace was about all this guy could handle.

Part of my responsibilities with CRU over the last few years in Russia was to help shepherd and give direction to other teams across Siberia. We had ministry teams working with students in several cities across the eastern part of Russia. Our team was in Irkutsk, another was in Vladivostok, one was in Ulan-Ude, and another was in Kyzyl, the capital of the Republic of Tuva. I visited each team at least once a year with another colleague. We usually spent several days with a team before returning home. Ulan- Ude is a city that was a distance of about a five-hour drive by car or about eight hours by train; it was the easiest one to visit.

The drive to Ulan-Ude is beautiful. There are vistas of several mountain ranges, and the southern coast of Lake Baikal is spectacular. With over 20 percent of all the fresh water on the planet, Lake Baikal is the largest lake in the world. It is crystal clear and more than a mile deep. It is also home to numerous indigenous species of fish and wildlife found nowhere else. One of those is the nerpa or freshwater seal. Lake Baikal also transforms into the best, smoothest road in all of Siberia every year. This road is open for only about three months each year. Siberia's bitter cold has drastic effects on the roads. The cold causes big cracks and potholes to form in any pavement, making paved roads the hardest to navigate.

The pavement lulls you into driving speeds that make it impossible to avoid the hazards. I have hit numerous holes that jarred my teeth and watched as other cars hit them and lost tires, wheels, and even brakes, shocks, and struts. This same cold, however, makes a road smoother than any racetrack. My speedometer never went above fifty-five miles per hour unless it was on Lake Baikal, and then I felt the exhilaration of opening it up to eighty or ninety. The winter route cut about ninety minutes off the trip to Ulan-Ude. The only challenge to this route came when visibility was low due to a winter storm. Then the occasional pine tree, cut and planted in the ice, marked the way across. Using a compass helped immensely, but the sightings of a tiny pine tree still brought great relief.

Vladivostok is two and a half days by train or a four-hour flight. The visits I made to our teams there were always welcome trips. It was still a cold, gray Soviet city, but it was on the Pacific Ocean, making it a stark contrast to Irkutsk and the interior of Siberia. Smelling the sea air and seeing the huge ships reminded me that there was another world I barely remembered. Another benefit to this trip was seafood—fresh, cheap, right-off-the-dock seafood. The favorite was Alaskan king crab legs, although they left off the "Alaskan." Once again, the Lord allowed me to eat like a king, this time on a missionary budget. I have some fond memories of times in Vladivostok.

Kyzyl in Tuva is the antithesis of Vladivostok. The Republic of Tuva is tucked along the northwest border of Mongolia and is home to the Tuvan people. They are closely related to Mongolians, but their language is quite different. Kyzyl is the capital of the republic and also the geographical center of the Asian continent. Asia is the largest continent, and Kyzyl is located exactly at the center of the continent; therefore, when a person stands at this monument, he or she cannot get farther from an ocean anywhere in the world. Literally, no one is farther from an ocean in the world than he or she is at that moment.

Hence, there is the antithesis of Vladivostok. There is no railroad connecting Tuva to the rest of Russia. The one road is difficult by car except in summer. Fortunately, a few flights connect Kyzyl to several cities in Siberia.

In spring 2003, I purchased tickets for a colleague and me to fly from Irkutsk to Kyzyl to visit our team there. This was my first trip to Tuva, and I was excited to encourage our team, meet some Tuvan students, and visit the geographical center of Asia. I was also excited to have a bird's eye view of the Sayan Mountain Range. This is one of the most rugged and beautiful ranges on earth. It is home to argali sheep, ibex, Asian grizzly bears, and places few humans have ever been. I managed to get a window seat for the two-hour flight to Kyzyl, and with good weather, I had an unobstructed view. It was breathtaking and a two-hour drink of water for my soul. There is always something therapeutic about spending time in the mountains or any place of beauty. I may have been tucked in a seat, peering through a one-foot-square window, but it was some serious therapy. I made a mental note to make sure to get a seat assignment that would give me a view looking in the other direction for the return flight.

The return flight, however, never happened, at least not the return flight to Irkutsk. After a three-day visit with the team and of course the photo op to stand on the geographical center of Asia, it was time to head home. A taxi dropped us off at the tiny airport. We probably could have walked, but it was only spring on the calendar. Spring hadn't yet arrived in Siberia and wouldn't for another month. Tuvan Airlines is more like a premature "Baby-Flot." With only a couple of prop planes in their fleet, it was too small to even be considered a true "Baby-Flot."

We strolled up to the only desk to check in but had trouble locating an agent to help us. Finally, a woman appeared from somewhere and asked what we wanted. I handed her my ticket as I said we were checking in. Without looking at my ticket, she asked, "Checking in for what?"

I answered, "The flight to Irkutsk?" but it was more like a question than giving her an answer. Since we were the only two people in the airport, I began to wonder whether I had the right day. Had I missed the date? Was I an entire day off or maybe just an hour off on my calculations? I rechecked the time and date on my watch, while the woman finally looked at my ticket. My colleague looked at her ticket too, and we concluded that we had the right day and time.

Without any facial expression whatsoever, the woman said, "There is no flight to Irkutsk; we stopped flying there." Of course, this response evoked some facial expressions from me.

"We just flew from Irkutsk three days ago, and this ticket says Irkutsk," I protested. By then I had grabbed the ticket back from her and was pointing to the word "Irkutsk." "From Kyzyl to Irkutsk," I repeated, my volume increasing slightly.

Her demeanor didn't change, and she repeated her matter-of-fact line again. "There are no more flights to Irkutsk."

I had bought a ticket for a flight that didn't exist! Who was the fool now? I stood there for a moment, unsure of what to do or say. Finally, I asked, "Why do I have a Tuvan air ticket to Irkutsk?"

A shrug was all she returned to me. I then tried to calm myself and asked whether there were any other flights leaving Kyzyl today. "No, none," she said without any compassion or remorse, "but tomorrow morning there is one to Krasnoyarsk."

Feeling defeated, I slowly allowed the white flag of surrender to be raised. I had nothing, and she knew it. My colleague and I stood there for a few minutes in the empty airport, wondering what we should do. The ticket agent still stood across the desk, looking at me like a boxer who had just won by unanimous decision. I now realized she had all the power—the power to sell a ticket, refund a ticket, or keep us in Kyzyl indefinitely.

I changed my attitude and tone. With a smile that lacked all integrity, I inquired about a refund. "Not here. Only the person who sold you the ticket can give you a refund" was her comeback.

I was now on the edge of losing my religion. If I cursed in English, would it make me feel better even if she couldn't understand it? *Don't go there. We still need a ticket unless we want the geographical center of Asia to become home*, I reasoned.

With my tail tucked firmly between my legs, we purchased tickets on the only flight out of Tuva for several days. This did give us one more day with our team and another chance to grin and bear it. The next day we were back at the airport, and this time there were other passengers checking in too. Good sign. The determination not to lose my cool was made easier by having a different ticket agent check us in. We were now on a flight to a destination that got us no closer to home, but at least it was a bigger city with more options. The train went through Krasnoyarsk, and a road connected it to Irkutsk, so with any luck, we might be able to get home.

Luck? What luck? We landed in Krasnoyarsk to find there were no flights to Irkutsk for another three days. We hailed a taxi from the airport to the train station. Hopefully door number two would be open. It was half open! There was only one ticket on the next train heading east. It was an overnight train and would leave later in the afternoon. I purchased the ticket and gave it to my colleague. The next train was two days away, and I was overdue to get home. Planes, trains, and now a taxi?

I said goodbye, wished her a good voyage on the train, and headed out of the station. My taxi driver was still sitting in his car, hoping to find another fare. He saw me and asked whether I needed his services again. It took me about two seconds to decide to test the limits of his taxi meter.

"Have you ever been to Irkutsk or Lake Baikal?" I asked him, acting more like a travel agent. He didn't get my drift. "How much to drive me to Irkutsk?" I clarified.

He shook his head and said, "That's a fourteen-hour drive."

I agreed with a nod and repeated, "How much?" We finally settled on $150 if he could bring his wife along, since she had never seen Lake Baikal. Done.

An hour later, we were on the road with his wife in the front seat, a tank full of gas, and a bag of road trip snacks. I was getting home one way or another, and the adventure of it left my immediate desire for anymore completely drained. I'm not sure whether there is an entry in the *Guinness Book of World Records* for longest taxi ride, but I'd like to think I have a shot at it.

Chapter 16

# The Wild West

Many of us described Russia in the 1990s as the Wild West. Over one hundred years earlier, this term referred to a lawless period in the western United States, but the comparisons with Russia were easy to draw. There was a feeling that "anything goes" or "the toughest guy wins." Most Russian laws were considered suggestions and seldom enforced. Corruption was rampant and permeated every level of society. Most business owners had three sets of books: the actual books, the "cooked books" to show the government officials, and the books they showed to those extorting them. There was a price everyone paid who wanted to conduct business in this new "free market" economy. With the collapse of Soviet communism came the possibility of free markets and capitalism, but without the moral and ethical foundation to support it, capitalism just became crime. Evidence was all around. It wasn't uncommon to see kiosks that had been torched. How are these metal, stand-alone buildings so susceptible to fire? They weren't. They were susceptible to Molotov cocktails.

Early one morning, a loud noise that rattled the windows in our flat woke us. At first, we didn't see anything outside and couldn't tell from which direction the sound had come. Then as emergency vehicles responded, we followed the lights and sirens to a building across the courtyard from ours, maybe one hundred yards away. Daylight revealed that a vertical line of windows had been all blown out and blackened from smoke. All the windows in the stairwell of the ten-story building were gone. It took several days, but we eventually learned that a grenade had been tossed into the stairwell when the son of a prominent businessman came home. The message was literally loud and clear to those who wanted to do business: pay to play. The only profession that was more dangerous than being a businessman in Russia during that time was being a politician. The life expectancy for both was rather short.

The lawlessness rarely touched us directly. Our apartment and garage were never broken into, and we felt like God was protecting us. During the eight years we lived in Irkutsk, we were the victims of only two crimes, but neither threatened our personal safety. The first cost me only what I had in my pockets. The other cost a bit more.

Our team lived in a section of town called Universitetski because of the proximity to several institutes. It was a logical area to live if you were working with students. Most of the apartment buildings had been built in the 1970s. They were mostly prefab concrete buildings that all looked identical. During construction, it was decided that if there were, for instance, fifty apartments in a building, then there was only a need for thirty telephone lines. Don't ask me what the thinking was behind this. When we had purchased our apartment at the end of our second year, a telephone line and number came with it. The phones were less than reliable and seemed to be out of service once or twice a week. Few people had cell phones at that point, so we, like everyone, just accepted it and were thankful for when the phones did work.

One of the keys we found to being content in Russia was summed up nicely with the phrase "high anticipation, low expectation." This applied to phone lines as well as many situations. These phone lines also brought us the dial-up-speed Internet we were all becoming more and more dependent on. The Internet speed did increase after a few years and with it the usage and perceived need for it. Since phone lines were often out, these became the grin-and-bear-it times or times to go to a friend's or teammate's flat if you really needed a phone.

We had an outage during our fifth year that seemed longer than most. I had been grinning and bearing it for two days and was beginning to lose hope. On day four, my grin was gone for good, and I headed to the phone company office in the city center. Once at the office, I waited in line to talk to a woman who had lost her grin too. I explained my situation to her in my fourth-grade-level Russian and was met again with the familiar blank stare that shouted, "I don't care."

Her first question to me was whether I had paid my bill. Of course, I had and had never been late. She checked her computer, which confirmed I had indeed paid every bill on time. She paused and leaned closer to the monitor. She squinted a little and said, "Oh you don't have a phone anymore."

I was confused. What did she mean? I still had a phone line coming into my apartment. It was still connected to the actual telephone sitting on our hall table, and I knew it didn't currently work. I tried inquiring again, and she shed a bit more light on my murky situation by adding, "We gave your telephone line to someone else."

What? Why? These were questions she couldn't answer. Or didn't want to. I wasn't sure. She pointed to a door across the room and said I would need to talk to the director about this matter further. I tried not to stomp as I made my way to the office door with "Director" written on it. The door was locked. I noticed the office hours posted a few feet away. I have never seen office hours before or after like this: Tuesdays, 10:00 a.m.–11:00 a.m. I kid you not. Who has office hours one hour a week? Since it was Wednesday, I knew I would be getting some exercise for the next week by walking to our teammate's flat to use the phone and Internet.

On Tuesday at 9:45 a.m., I reentered the phone company's lobby doors. We had made it through eleven days without a phone line, and I was hopeful that the situation would change today. My hope slowly slipped away as I realized there were eight people ahead of me in line to see the director. I still waited and held my place, but after the first person took twenty-five minutes with the director, I saw myself getting good exercise for at least another week.

The following Tuesday, with less patience than the week before, I walked through the doors, this time at eight a.m. I was first in line. It was now the one who watched others make the hopeless walk across the lobby to find a line of people in front of them.

The director unlocked his door a few minutes after 10:00 a.m., which was my cue to head in. I sat in front of his big desk, and we exchanged greetings. I told him who I was and described my situation, and I tried my best to do so without any animosity. He too checked his computer and then, in the same manner as his assistant two weeks before, stated, "We gave your phone line to someone else." I asked why. He said that three years earlier, I had never put the phone in my name after purchasing the apartment. I protested and showed him a phone bill with my name on it, to which he didn't have a reply. He then added that there were a limited number of phone lines in each building, and he could do nothing unless someone moved or died; then that person's line would become available to the next person.

I was angry, yet we needed a phone. He then added that I had never paid to have the name changed. I asked how long it might be before another phone line would be available. He shrugged and said maybe six months or longer. I was now playing the game against my will but playing nonetheless. I asked whether there was a way to expedite getting a new line. He said yes, there was, but he still didn't offer a solution. I finally asked the critical question. "How much does it cost for expedited service?" Before he could answer, I reached for my wallet. I pulled out a wad of rubles, worth over $100. Apparently, that was how much it cost for expedited service. I wasn't sure whether I fully trusted him, but he was true to his word. He expedited the service because by the time I arrived home, we had a phone again.

The other time I felt like a victim was when our car was stolen. We had been warned about this tactic. Someone takes something that belongs to you and then holds it for ransom. We were back in the

States on furlough and had left our Toyota Rav 4 for our teammates to use. Car theft was common, so we had taken many precautions to hopefully prevent it from happening. We had installed a car alarm, which disabled the ignition when it was activated. Our garage was nearly bombproof, made of two-foot-thick concrete walls and huge steel doors. The car was vulnerable only when it was parked on the street. Our teammates had driven it into the city center and parked it on what most would consider the busiest street in the city. Even with all this security, they came back to find our car gone.

Obviously, the alarm system wasn't as foolproof as we had thought. Sure enough, after several days, someone called our apartment phone. They had acquired our phone number from the vehicle's registration. We had heard of these kinds of car thefts happening all over the city. After the theft, someone contacts the owner and offers a "half-off sale." For half of the value, they will return the car. I wondered why the police weren't able to catch some of these thieves. It seemed like a sting would be able to nab some of these guys, unless the police were in on the scheme somehow. After all, somehow these criminals were able to use only the registration and obtain phone numbers.

Anyway, the crooks contacted our teammates and asked for $5,000. By now, we had been informed about the car and made it clear we didn't want to encourage this or have someone negotiate with these people. The thought of taking five grand in cash to pay known thieves didn't seem wise. I told our teammate to just hang up on them when they called again. Eventually, the calls quit coming in, and we were out a $10,000 Toyota Rav 4. Fortunately, our teammates helped us to replace the car when we returned from furlough.

We rolled the dice and purchased another car not long after our return. A few months later, I discovered just how high up the chain this corruption went. I went back to see my friend Alexander at the Fish and Wildlife Department. It was good to see him and bring him a few gifts from the United States. I finally got around to telling him about our car being stolen. He immediately asked whether the thieves had contacted us for payment. I said they had and told him how things had gone from there. He shook his head and told me to always call him when anything like this happened.

To be honest, I had never thought about burdening him with something like this. After all, he was in the business of deputizing me as a weekend warrior game warden, not solving grand theft auto cases. But he insisted that I let him know if the situation ever happened again and explained how it would turn out differently. He said, "When the thieves contact you, get their phone number and tell them you will call them back with your answer." Then I should bring him the number, and he would give it to some friends. These are some of the same friends who now enjoy cow elk calls. "My friends," he explained, "will call the thieves back and let them know whose car they stole. When they find out, they will not only return your car to you but also return it with an apology."

I remember trying to connect all the dots in this criminal scheme. My head hurt from trying, and this confirmed that I really didn't want to know how high up corruption like this went. I was convinced that I would call Alexander if anything like this happened again and that a few more elk calls would be a good investment. Fortunately, nothing like this happened again during our adventures in the Wild West of eastern Russia.

Chapter 17

# Door or Barf Bag

In Romans 15:20, Paul told his readers that his ambition had always been to preach the good news where Christ wasn't yet known. I remember underlining this verse in my Bible as a college student and praying that God would use me likewise. Later, I wrote this verse on maps of Siberia as I began praying for the people groups across that vast land. One of the reasons for moving to and ministering in this part of Russia was the proximity of some of these people. Four indigenous people groups near Irkutsk were considered unreached: Buryats, Tuvans, Yakuts, and Evenks. These people were also represented among the student population in Irkutsk. Together, with our team, it had become our ambition to see God's love and forgiveness embraced by these students as well as the Russian population. We often prayed for opportunities to meet students from these ethnic backgrounds. Slowly but surely, we did.

Many of the students involved with the student ministry also became members and leaders in a small church plant. Dear friends of ours had planted a church in Irkutsk we were a part of from the beginning. It was such a blessing to watch as God matured some of these students to the point of reproducing their faith. The more this happened, the more apparent it became that it was time to get out of their way. Our strategy had always been to work ourselves out of a job and turn the ministry over to nationals as soon as possible, but it was also hard to know when that timing was right. Plus, it was hard to let go of something we had helped birth.

We were in our eighth year in Irkutsk. Our team had grown, and we now had four national staff working with us as well as numerous volunteers. There were teams in three other cities as well and a lot of momentum. During this time, I was studying the book of Romans again and was back in chapter 15. This time it wasn't verse 20 that caught my eye but verses 23 and 24 instead. Paul wrote that "there is no more place for me to work in these regions." I had a hard time digesting this at first. How could Paul say there were no more places for him to work? It wasn't like the gospel had been heard by every person in the Middle East or that churches had been planted in every city and town. So why would he write that?

Then it hit me. Paul knew that *his* part was done. He had been the starter, the one to get things going, and then it was time to move on. This ambition drove him to take the gospel to other places and people who hadn't heard yet. It was then that I sensed the Lord saying to me, "Spencer, you have been my door here." Together with a team that had helped pioneer this ministry, we had opened the door for others to walk through. Now it was time to step aside and let others lead the ministry. It was time to let others, with better language skills and different gifts, take the reins.

As the Lord often does, he began to confirm this in another way too. During the winter of 2005 on a Sunday morning, we enjoyed watching many of the students lead the worship service at our church. Some were helping with worship music, one was delivering the sermon, and others were serving communion. That morning, at least one student from each of these four people groups was involved in the service. Suddenly, what I was witnessing dawned on me. I began to cry as I thought about the significance of this moment. God had done what we asked of him. He had used us and brought to fruition something he had put on our hearts years before. At least part of the dream had become a reality. I knew then and there that it was time to leave Irkutsk, and God was calling Annette and me somewhere else. I just didn't have a clue where that was.

We prayed a lot about this decision over the next few months, asking God to make it clear where he wanted us next. We started to realize that sometimes God's calling on our lives comes in chapters. It seemed that he was turning a page, and with it a new chapter was starting. But we had examples from others that seemed like God had called them to the same place and people for most of their lives. My great aunt, Margaret Nicholl Laird, spent nearly fifty years in the Central African Republic as a missionary, first as a single woman, then as a wife and mother, and later as a widow. I wrestled with this for a few months and finally concluded that God had called me first and foremost to himself. My primary calling was into a relationship with God, then to make him known to others. I remember hearing somewhere that there were only two kinds of people in the world: missionaries and those who need them. My life belonged to God; it was no longer mine. He had bought me and paid for me. Maybe geography wasn't that important.

CRU offered us jobs in a few places with other teams, but none of those felt like a good fit. During our time in Russia, we had seen many missionaries come and go. We watched as many of those left the field for the wrong reasons. God is always sovereign and can direct us through difficulties as well as other methods. Instead of God calling some of these folks to other fields, many were forced to leave their ministries because of poor health or poor team relationships. Some others, it seemed, were never really in the place where God had wanted them from the start. Some were called, and some just showed up. Still others had come to the mission field—not because God led them but because they were running from something else. We were burdened for these folks and thought God may want us involved with his servants in situations like these. Maybe God wanted to use our experience to help others continue in their ministry. But where and how?

God's timing is perfect. Around this same time, we found out my brother Matt was taking steps to start a ministry in Colorado and using the family guest ranch as the setting. It wasn't clear yet what it was going to be, but it was a chance to start something again. Not only was there another door open, but with it came a better understanding of how God had made me. I was learning that I loved to start things, open doors, and ask God to use us. This was another blank canvas, and if we let him, God would start with the broad strokes and show us what he wanted.

Annette and I felt that part of this new ministry would be to pastors and missionaries and their families. We wanted to help those already in full-time vocational ministry to continue from a place of fullness. It seemed like so many weren't thriving and ministering only from their reserves. Cups weren't overflowing. So, in the summer of 2005, our family moved to Ohio City, Colorado, to the Big Horn Guest Ranch, where I had grown up. Since the ministry would be based there, the name became Big Horn Ministries (BHM). We were blessed to have so many friends and partners believe in us and support us as we moved out into these uncharted waters.

Starting a nonprofit was only one of the challenges. The biggest challenge was keeping our hands off this new ministry enough so God could make it what he wanted. This was a process that took several years. Big Horn Ministries became a patchwork of different opportunities to work with a variety of people.

One of the new opportunities was helping bring refreshment and the gospel through outdoor activities. Hunting, fishing, and hiking were wonderful ways to rub shoulders with people. We began partnering with a foundation that helped youth with a life-threatening disease or illness go on dream hunts. As of 2021, we have had the privilege of hosting fifteen kids on deer and elk hunts. God was opening new doors, but the main focus of BHM would still be pastors and missionaries. The mountains of Colorado are a perfect place for those who need a place to unplug and recharge their batteries. We also began to see why God would have us leave an established organization to start something new. We found that if we didn't have any official ties to churches, denominations, or mission organizations, there was a greater chance of openness and vulnerability.

We began learning how to be safe people and offer a safe place for leaders to unpack their journey in ministry. We soon realized we had a spiritual gift that was quite useful, that of being "barf bags." Sometimes folks just need to throw up before they can feel better and move on. God knew what some of his children needed and how he wanted to use us. Once again, the real adventure was simply walking with God and letting him lead. Looking back now, we would have never written the script God wrote. But the more he wrote and unfolded to us, the more doors were opened. It became obvious that he had equipped us for this chapter too.

Chapter 18

# Missionary meets 007

Transition to life back in the United States wasn't easy for our family. There were multiple challenges to overcome before we would feel settled. Andrew and Kate were third-culture kids, which gave them both a different outlook than we had. Because they had grown up in Russia, part of them identified with that culture but not fully. They each had US passports but weren't culturally American.

Along with other kids who grew up in a host culture different from that of their parents, they develop their own culture or a third culture. With this comes a less ethnocentric worldview but also a few challenges. It's hard for many third-culture kids to find their place and fit in. God helped both of our kids to navigate this road, and eventually they adjusted well.

With every challenge comes an opportunity. First, there is an opportunity to trust God. But also the lessons God teaches through these challenges are never just for us alone. God will use those lessons to teach others through us. All four of us encountered challenges as a result of living and ministering overseas and then moving back to the States. God has given both Andrew and Kate unique opportunities to grow and serve others as a result of their experiences. And the challenges life and transition brought to Annette and me God wanted to use too. At first, it came through hosting people at the ranch in

Colorado. Then God began to open doors for us to bless others by going to them.

The first door opened after a couple of years of living back in Colorado. Annette and I weren't strangers to doing international travel or being in difficult situations. So invitations began to arrive for us to be involved in some short-term ministry opportunities. At the time we had friends living and working in Muslim North Africa. They are experienced and seasoned veterans and understood well what it would take to reach Muslims. Like us in Russia, one of the challenges for them was how to get and keep visas to live in this country.

Because of this issue, there were some risks not worth taking. They wanted to distribute portions of the New Testament as well as audio recordings of the Bible and DVDs. Of course, giving away anything like this was illegal in this country. Our friends shared the need with us and that they just couldn't risk losing their visas if they were caught. So would we be willing to come over and distribute materials, knowing that if we were caught, we would likely be stripped of our tourist visas, escorted to the airport, and not be allowed back into this country? The proposed plan was for us, together with another couple, to pose as tourists and travel around the country in a rental car. Along the way, we would stuff mailboxes at night with materials and hand them out at our discretion.

After much prayer, Annette and I came to the conclusion that this was a door God wanted us to walk through. We knew whom we wanted to join us. It took our friends Brad and Becky less time than we did to say yes. You can do anything if you have friends with you. Plans were made, tickets were purchased, and bags were packed; soon we were off to North Africa. Part of the preparations was to get many people praying with us for this trip. We didn't take this assignment lightly and knew there were risks. But then again, there are risks we face every day. Upon arrival, we would make contact with our missionary friends only once. Then we would pack the trunk of the rental car full of materials and, with our personal bags on our laps, begin a loop tour through the country.

We were also given a cell phone with two SIM cards. The first one was to call home and talk to the kids, parents, and so forth. The second SIM card was to be used only if we were busted. If the authorities caught us, then we would insert the unused SIM card into the cell phone, call our missionary friends with a two-word message, and then hang up. "We're dirty" meant we would have no more contact with them and would likely be tossed out of the country. It had the feel of mission work meets 007. It had the feel of adventure.

Things went according to plan as we arrived and rented the car. The four of us had packed light to spare the trunk space for materials. The literature, CDs, and DVDs had been smuggled into the country earlier, and our friends had stored them. There was a significant risk, but there was potential reward too. The reward would be giving those who were thirsty to know God an opportunity to start a relationship with him. The reward for us would come with the understanding that we had played a role in building God's kingdom while having our faith muscles stretched.

The first few nights we were in bigger towns, staying at tourist hotels. We would wait until just after midnight and then make our move. Two of us would stand guard, and two of us would stuff mailboxes in apartment buildings. Then we would move to another building and repeat this process. We would get to bed late and sleep late the next morning before visiting another tourist site. We looked the part of tourists and took a lot of photos.

Things went smoothly for nearly a week. We had distributed much of the trunkful and had only one day and one tourist destination left on our agenda. We planned to give out a few more sets of materials as we met people and asked for directions to the next town. As they shared where we should be driving next, we often thanked them and offered a small gift of thanks. Sometimes we asked for directions every few hundred yards just to have another opportunity to give away more materials. After visiting our last tourist destination, we planned to head back to the airport in the capital. We still had several boxes of the precious literature left and hoped to return them to our friends before our departure. After all, it had taken a lot of work to translate, print, or record and smuggle these resources into the country, so we didn't want them to be wasted or fall into the wrong hands. I had given away one last gift of materials and DVDs to a young man, who thanked me.

We pointed the car toward the capital and started our three-hour drive. Twenty miles toward the next town, we were stopped. An unmarked police car was waiting at a makeshift roadblock. I pulled over and waited, unsure whether this was a routine traffic stop or something worse. I rolled down the window, and in broken English, the man said he needed our passports. He wasn't angry; he was dressed in a coat and tie and didn't at all look like a traffic cop. He showed me his badge, and I knew this wasn't traffic related. Reluctantly, we handed him our passports, knowing that without them we were going nowhere. He said we needed to follow him over to his office. We took turns praying as we followed him on the short ride to the police station.

Situations like this are quite revealing. They quickly cut through self-confidence and misplaced trust. My first thought was, *What will I say, and how should I say it?* Control is such an illusion. The best positioning and manipulating reveal only how deeply I think I can maneuver in these situations. In reality, they should force me to lean into God and trust him as my Father who knows and sees all. Eventually, in my heart, I said, "We're in trouble, and only you, God, will get us out of this mess." I suddenly just wanted to be home with my kids. I knew Annette did too, as did Brad and Becky.

At first, we sat in the car and waited for instructions. Then it became apparent we weren't going into the station but would be questioned outside. Our passports disappeared inside with the plainclothes cop. A few minutes later, he returned without our documents and started in on us with his routine.

Where did we get the materials? Why did we give them to the young man in the last town? Who translated them into the local language? Because his English wasn't good, this interview was taking longer than it should have. The time lapse gave us a chance to come up with suitable answers, some of which we had already rehearsed in preparation for this event. In unison, we said we didn't know who had translated the materials. I then added that we had heard that this culture was a gift-giving culture and very friendly, so we wanted to have a gift to give a new friend. This seemed to appease him for the time being. Then the "bad cop" approached us.

The routine was obvious, even through the language obstacle. The method of good cop and bad cop was evidently practiced outside the United States and probably the world over. The bad cop was in uniform and seemed to have his part down. Without a smile or introduction to us, he started with a statement that shook us. He said matter-of-factly that people caught handing out this kind of religious material could be shot. I remember gulping at the word *shot*. Oh, how I wanted to snuggle with my kids right then. Bad cop waited for that word to sink in.

I can't remember exactly what we said in response, but it was something about not wanting to offend them or their culture; but we had just wanted to share a gift with this young man. Good cop returned with a smile and lowered our blood pressure some. He seemed to know of other places we had visited in the country. It was a good thing we had visited the main tourist sites and had that card to play. But it wouldn't take them long to find others who had also received "gifts" from us as well.

There was also one big elephant that had yet to be noticed. There were still two boxes of materials in the trunk. If they searched our rental car, the discovery would be a game-changer. Then they would see for themselves just what we were giving away and have evidence in their hands. Our silent prayer focus shifted to one Brother Andrew had often prayed. Andrew van der Bijl was a missionary who smuggled Bibles into communist countries starting in the 1950s. His prayer was often, "God, you made blind eyes see. Now I pray you to make seeing eyes blind." We added, "Please don't let them have the thought to search the car or open the trunk!" God answered that prayer for over two hours. Not one cop asked to look in the trunk.

Good and bad cop went back and forth for some time. In one lecture, bad cop stated that if we ever wanted to give a gift to someone in this country, we needed his permission first. Even though he was the cop in charge of a small town, I think this made him feel more important to make statements like that. Trying to break the tension, Annette and Becky finally pulled out small photo albums of our families. They passed around photos of our children while talking about how much they missed their babies. Many of the uniform cops seemed to be moved and really wanted to see the photos. Photos of the children and families led to showing them photos of where we lived in Colorado and even postcards of mountains and wildlife. It's harder to be tough and intimidating when looking at photos of children and beautiful scenery.

We could see this method was having some effect, so we even offered some of the postcards as gifts to the officers. About then, Mr. Grumpy Pants returned, and we realized we had already disobeyed his last order never to give a gift unless we had his permission first. We began to jokingly ask for the postcards back, since we hadn't sought his permission. Fortunately, he also thought this was humorous, as did all the others. The tension was ratcheted down a little more, and each of the cops accepted a postcard.

Good cop came out of the station soon after this and walked slowly down the steps toward what was now a crowd of cops around the four of us. Even though he was the nice cop, clearly he was also in charge. In his hand, I could see our passports and felt the weight again of what they represented. He had one more tongue lashing for us that made him feel good given the audience, then handed our passports back to us. It was over. We were free to go, but the message was reiterated: "Never give a gift without permission first!"

There was a huge collective sigh of relief once we were on the road again, along with a lot of thanksgiving. Suddenly, we had the urge to get rid of the remaining materials. But first, we needed to get some miles between us and this town. We also needed to use that other SIM card. We were fearful of getting our friends in trouble and jeopardizing their work. Guilt by association would be instant if the wrong people knew. I made the call. As soon as I heard his voice, I softly said the two words, "We're dirty," and hung up. The plan we had devised was well thought out and would have worked fine if we hadn't had any materials left. Oops.

Maybe the plan wasn't thought out very well. The cell phone rang a few minutes later. Our friend wanted to know briefly whether we had materials left. He was willing to risk guilt by association to retrieve the materials, which he did. We prayed for God's protection over this mission team and that the rest of the materials would reach the hands and hearts that needed them. Our friend also informed us that our passports would be flagged and listed by the government so we wouldn't be allowed to reenter this country for many years. Oh well, I likened it to "better to have loved and lost than never loved at all." Better to have gone, been flagged, and not allowed to return than not to have gone at all. When we arrived home, we couldn't stop hugging our kids.

This adventure had caused me to drill down deeper into my faith. I knew my faith muscle would get more exercise and not just through work but also through hobbies.

Chapter 19

# Fear can lead to faith

One of the best things about any adventure is the unexpected, the things you never see coming. Any time we leave our comfort zones, the possibility exists. For me, this usually happens when I am in the outdoors. I am not uncomfortable in the outdoors, but I have chosen to leave a more controlled environment, the confines of home, or being with other people, which brings a sense of security. Time alone in the outdoors allows the opportunity for the unexpected to happen on a more consistent basis. I cannot count the times I have ventured out with some plan to hike or hunt, and I find myself in a situation that wasn't at all what I set out to do. The primary adventure unfolds into secondary and tertiary experiences, many of which you would have never agreed to if you had known ahead of time what was going to happen, like steps leading you farther away from security and deeper into danger. Of course, the first step or two were of your own choosing.

As I have learned to walk with God at my side, I realize that many of these experiences have accompanying lessons. Some of these lessons are of a smaller variety; they are for the moment or for the next time you find yourself in a similar situation and can draw from the knowledge acquired. Others seem to yield more of the "life" lessons. The kind you need to keep in long-term memory and will be applicable many times in the future. Many of my outings, which have been interrupted with the unexpected, have pushed me deeper into faith. Faith is a lesson that is never fully learned. Faith can grow if we exercise it, but we never graduate.

This growing faith has bolstered my courage in many situations where I was tempted to turn back and not continue. In hindsight, discretion may have been the better part of valor on a few of those outings. But discretion and the retreat it advises often come more like a tortoise than a hare. And before you know it, you are in deeper than you bargained for. These deeper or frightening situations have thankfully yielded a few outcomes. One, they resulted in an adventure story that is difficult not to share. The other outcome is hopefully a lesson taken to heart.

There are many dangers one can face in the mountains of Colorado. Driving snowy mountain roads, avalanches, and mishaps while recreating are among the most common. Encounters with wild animals that put people at risk are uncommon. Many visitors and tourists I have hosted over the years ask about the dangers of several animals, mainly carnivores. While the possibility exists of being trampled by a moose or elk, it is extremely rare. At the same time, it is rare to have a dangerous encounter with one of our carnivores. Lions and bears are the only large mammals that fit this category, and bears are actually omnivores, eating vegetation along with meat. Still, the occasional news story of a mountain lion attack or bear attack gets your attention. Because they often kill to eat and have large canines, the questions are typically about these animals.

Black bears are plentiful in the area where we live. It isn't uncommon to see them in summer months, crossing a road or looking for an easy meal. It is always a treat to get to see one of them and appreciate its power and beauty from a distance. When asked about the danger associated with bears, my dad always had the same answer. "Black bears are more scared of you than you are of them." I adopted this response from Dad and repeated it often. We have watched and hunted bears for decades, and it was true; most bears ran off immediately if they sensed a human presence. There are exceptions to this rule, especially as humans continue to encroach into more and more bear habitats or carelessly leave trash and food bears can consume. I still liked Dad's answer, though, and used it right up until the time I found it wasn't always true.

I was archery elk hunting in southern Colorado in a new area with my good friends Wayne and Jon. They had hunted this area for years, but it was new to me. The elk population was healthy, and anticipation of the first morning's hunt was high. Before dark, we went our separate ways and agreed to touch base with each other by text about midmorning. After about a mile ascent, I was near the top of a ridge and close to elk. I could hear several bulls bugling and knew I would soon have a close encounter with one or more of them. I managed to get close to one mature bull but failed to make the shot. I followed in the direction the bull had run, hoping to get another chance. After a few hundred yards, I knew it was useless. I had blown my chance, so I sat to catch my breath and consider what had gone wrong with my shot. I was now on top of a knob, which gave me a good vantage point. I could see down into the sparse trees dotting the steep slope below me. Pulling off my backpack, I grabbed my water bottle and chugged. Leaning against my pack, I let go of the adrenaline rush, breathed deeply, and relaxed.

Then at the edge of some of the thicker spruce trees, I caught some movement. Something big had moved below me about eighty yards. I grabbed my Matthews compound bow and knocked an arrow. The surge of adrenaline was back. I knew I might not have long since the wind was blowing from me in that direction. Elk, like most mammals, have an incredible sense of smell and can detect a hunter's presence with the slightest whiff.

I positioned myself and waited. Within a few seconds, I caught movement again. It was quartering to my right and moving uphill. It was still too far off and in too much cover to see whether it was the same bull. Another ten yards in the direction it was moving would afford me a better glimpse. As it emerged into the opening, I disappointedly realized the brown hair belonged to a medium-sized black bear. He had a beautiful dark-brown color, and his coat showed the signs of thickening as the weather turned colder. With his head down, he slowly waddled up the steep slope. He didn't seem to be in a hurry; he was just meandering in search of a food source as bears often do. He walked with a slow, steady gait.

In the fall, bears need to gain weight in preparation for hibernation and often look filled out and robust. This bear, however, looked thin for the time of year, even emaciated. I could see his hip bones and thought it strange. A bear's hind quarters typically are where a lot of fat is built up and stored for the long winter. A big butt is normal for fall, and I thought this bear needed to eat a lot over the next few weeks. The other thought I remember having was, "Why hasn't this bear smelled me and dashed away?"

I let him walk a few more steps until he was about forty yards away, still well below me but close enough. I was sure he had caught my scent by this time, but I wanted to make my presence heard also. So I let out a "Hey, bear!" spoken louder than necessary. I have done this on a few occasions when in similar situations and thoroughly enjoyed watching as the bear retreats at lightning speed. But this time there was no reaction at all from the bruin. He didn't even lift his head and look in my direction.

Had I found my first deaf bear that also lacked olfactory senses? What were the odds of that? Two more steps in the same direction, and he turned 90 degrees like a soldier doing drills. Now he was quartering to my left but still moving uphill and closing the distance between us. His gait seemed slower and even methodical now. I repeated my alarm phrase again and turned up the volume. "Hey, bear!" No reaction from him, just the slow rhythmic bobbing from his walk.

It was time to make myself visible too. Because I was wearing camouflage and sitting still, I wasn't easily detectable. I slowly rose to my feet, waved my hands, with the bow clutched in my left one, and yelled again. It was then that I knew this wasn't like any bear I had been around before. He didn't even acknowledge my presence. The only thing different this time was that I could see his left eye fixed on me even though his head never turned. This also meant he was close enough for me to see his eyeball, about twenty-five yards. Now I was fairly certain he had smelled, heard, and seen me; and yet he continued his approach. I decided it was time to up the ante. Without taking my eyes off him, I reached down and grabbed a baseball-sized rock. It was time for him to "feel" me too.

It is about this point, when I have told this story, that someone asks, "Why didn't you just shoot him?" I was elk hunting, not bear hunting. I didn't have a bear license. And in previous encounters, the bear had always spooked and run. So, I was determined *not* to shoot him ... yet. My first pitch was a ball. It was low and away and landed in front of his face a few feet. He jumped but then instantly continued on his path. The second rock I blindly searched for and found was more softball sized. It was delivered overhand with a little less velocity. This pitch was also low but right over the plate, directly in line with him. The one hopper hit the ground and bounced perfectly, smacking his shoulder. Each of my throws was accompanied with an emphatic "Get outta here!"

I was now out of senses. Feeling was the last one, and he had definitely felt it. Two more steps on the same route, and he did the 90-degree turn again, this time to his left. He was stalking me, no doubt about it. He was a determined hunter, and I was the hunted. Thankfully, my bow was still in my left hand and an arrow was knocked on the string and in its place on the rest. I attached my release to the string and pulled slowly back to my full-draw position. The bear was now about twelve to fifteen yards and still quartering toward me. I remember thinking that if he turned and faced me now, I would have lost my chance to shoot him in the vitals. Decision time! With my twenty-yard pin settled in just behind his left shoulder, I paused and spoke loudly, "One more step, and I will unzip you, bear!" This didn't seem to deter him any more than my previous attempts. Thankfully his next two steps were at the same pace and direction.

As promised, I let the arrow fly and watched it disappear on target. This time the bear received the memo and reacted as I had hoped all along. He spun around 180 degrees and bolted. Finally, he did something I expected and disappeared down the slope and into the spruces. Even with a fatal hit from my arrow, had he bolted toward me, someone else might be writing this. I sunk down next to my pack again, thoughts reeling. I was amazingly calm during this unexpected encounter.

When I relive or share the experience, I am more rattled than I was during it. I'm not sure whether I could make that shot again or even pull the bow back when I think about it. But in the heat of the moment, I had a surreal calmness. After a few minutes, the calm was gone. I began shaking like it was 20 below zero, and I had trouble catching my breath. *Slow down the breathing and pray!* Knowing the immediate danger had passed, I was eventually able to gather my thoughts and breathe deeply.

I knew I needed to do several things. First, contact Wayne and Jon and see whether they were close enough to head my way. Second, I knew I needed to track the bear and make sure it was dead. This was both for my sake and for the bear's. I fumbled through a text, hoping my companions had a cell signal and could find me.

Slowly I gathered myself as the shakes dissipated. I shouldered my pack, knocked another arrow with its razor-sharp broadhead, and started tracking. I stopped right where the bear had last stood and looked back at my post. There was nowhere to run or hide. I knew I had done the only thing I could, and I probably had waited too long to do that. I found my arrow after a short search. It had done its job, covered in blood. I found his trail and followed it into the thicker trees. Cautiously, I tiptoed around each tree and over a few logs, visually combing the forest floor as I went. Another twenty yards revealed what I desperately wanted to see: the bear sprawled out in a lifeless position, with two feet up in the air. He was dead, and the short nightmare was over.

There was something wrong with this bear. It wasn't just the way he acted. It was also the condition of his body that made me wary. His haunches and butt were sunken in, which left his hips protruding, evident even through his thick coat. His head seemed too large for his scrawny body, making me think he had been starving.

Wayne and Jon found me about an hour later. I was so happy to have someone to spill my guts to. I needed to tell someone what I had just been through. They both shook their heads, and together we thanked God for his protection. Knowing this bear wasn't healthy, we decided not to take the meat. In fact, no one wanted to touch it. It seemed like a waste but also not worth the risk of consuming meat that might be diseased.

I had started that day planning to pursue elk. Within a few hours, I had been the one being pursued. The adventure had unfolded nothing like I had planned. I was thankful I had a story to share. But I also knew there were lessons to be learned from all this. The immediate lesson was to start packing a handgun during my archery hunts. Who knows? Maybe a few shots fired in the air would have changed this bear's mind. And if not, I would have one more line of defense. The other obvious lesson became a vow. I will never repeat Dad's soothing advice again. The lesson that had become more lasting was the reminder that God is always present. And his presence brings peace.

It wasn't long before another carnivorous encounter gave me the opportunity to practice God's presence and accompanying peace again. The mountain lion population around Ohio City has steadily increased since I was young. I remember what a big deal it was to see a lion track in the snow when I was growing up. My dad had seen only one lion in all his years of hunting the area. I had never seen one until moving back from Russia in 2005. Since that time, I have seen twenty-one lions. The population has grown. After a snowfall, it isn't uncommon to cut at least one lion track on the roads within several miles of our house. I had been fortunate enough to harvest my first lion in 2007. A few years later, I was out looking for another one and to save some deer. Mature lions often make a kill every week or so. One lion may kill fifty-plus elk or deer a year. Because man is the only predator for mountain lions, hunting is the main management tool to control the population. I did my part and purchased a lion tag every year.

I had just let our black lab, Darcy, out of the truck to go for a run. She loved running on the snow-covered dirt roads. As I slowed down to let her catch up one evening, I noticed a lion track in the middle of the road, headed south toward Ohio City. With the dog in the car, I drove slowly and followed the tracks about two miles until they crossed over the snowbank and headed toward a cabin. My friend Randy was in Arizona for the winter, and I knew no one was home. His driveway hadn't been plowed, and the snow was two feet deep. The only tracks belonged to the lion. It was nearing dark, so I headed home and decided to come back at first light the next morning. I texted Randy that night to get permission to pursue the lion around his cabin.

I sipped a cup of coffee while I waited for the morning light. There is often a rifle in my truck during the winter months, along with a box of ammunition and a headlamp. Armed with these necessities and warm clothes, I headed out into the brisk, minus-25-degree morning. Wading through the snow warmed me as I followed the tracks toward the cabin. I found more tracks as I approached Randy's elevated hot tub about twenty yards from the cabin. It doesn't take a savvy outdoorsman to track something through two feet of snow. But the closer I got to the cabin, the more tracks there were. They were going in different directions, and it was apparent this lion had spent more time at Randy's cabin than he had over the last month.

The snow had slid off the roof all around the cabin and the front deck, making a four-foot ice mound. I stumbled over it and onto the deck. Near the steps was a hole in the snow, about eighteen inches in diameter, leading under the deck. Clearly, this lion liked the shelter of the deck and had crawled in and out numerous times. The deck was about twenty feet long and wrapped around the cabin. I decided to investigate the other side of the cabin. The snow and ice had created a one-way-in-and-out hideaway. As I walked back near the steps and the one entrance, I suddenly felt something under my feet. The space between the floor joists and the ground was only about one foot. Something was under the deck, and as it moved, it was hitting the joists. I ran to the other end of the L-shaped deck and again confirmed there was no other escape route. It was back to the entrance and decision time.

I was now convinced the lion was under the deck with only one way out. I could wait, but it was 25 degrees below zero. *How badly do I want to harvest another lion?* I reasoned. *It would save a lot of deer, and I am cold.* Lion hunting is an adventure, but this was now beyond the primary adventure, and I was headed for the unexpected secondary.

Still lacking some of the discretion valor often needs, I decided to slowly kneel and check for my quarry. I opened the bolt on my Remington Model 700 and made sure a round had been chambered. I slid the safety forward to "fire" and dropped into a prone position. With the rifle in one hand, I army-crawled near the entrance. Barrel first, I nudged forward. With snow piled up, it was dark under the deck. I couldn't see anything. Not good!

I reached for the headlamp in my coat pocket. The LED light was bright but didn't have a beam that went beyond three feet. I figured I just needed to see eyeballs, and they would be my target. I inched forward, hoping I wouldn't blow a hole in Randy's deck. My search with the headlamp yielded nothing: no eyes, no cat, nothing.

I pulled myself out of the hole and brushed the snow off. It's a funny thing how little we use the common sense God gave us when adrenaline pulses through our bodies. Adrenaline is quite addictive. That's why there are adventure junkies and many extreme sports. Once we have had a taste, it's hard not to go back for another round. I wasn't enduring this cold for nothing, but I also wasn't excited to poke my head back into the only escape route for this lion.

I walked back to the end of the deck again to make sure the lion hadn't dug another hole in the snow to escape. Standing on the far end of the deck, I felt and now heard the commotion under my feet again. This time the lion was headed in the other direction and toward the hole my head had just been plugging. I turned just in time to see the lion squeeze through the opening and bound away.

I couldn't risk taking a shot because Randy's neighbor lives about two hundred yards up the valley, the direction the lion was heading. I fully expected the lion to run up the pine-covered mountain out of sight and danger, or at least to climb a tree as they often do when scared. This lion did neither; he went straight to the next building.

What took the lion only a few seconds to cover took me a few minutes. Running for two hundred yards through two feet of snow will cause heavy breathing. That kind of breathing is hard on the lungs when it is 25 degrees below. I slowed to a jog as I approached the neighbor's shed. I felt as if I had just smoked a pack of Camels without filters. Coughing and spitting, I had trouble catching my breath. I felt something deep in my chest I hadn't felt in years. I knew then that I had frostbitten lungs. I hacked for several months after that morning. Trying to cough into my coat, I followed the tracks around the shed and to the entrance. One set of tracks going in, no tracks going out. Apparently, this lion preferred shelters built by humans.

Curtis is a contractor and quite talented. He had built homes for decades and had lots of snow-covered building materials inside and out. The shed was used for storage and had no door in the four-foot opening. I could see a lawnmower, motorcycle, boat seats, and tires stacked around but no lion. Inside the shed was another smaller, closet-sized room with a door open. Darkness cloaked most of the interior even without a door on the twelve-by-twelve-foot shed. Knowing how well my headlamp had worked in the last dark lion shelter, I opted for something with more lumens. It was time to see whether Curtis was awake and had a real flashlight.

I quickly made my way past his shop and toward his house. I was met by his wife, who was coming out the door and taking their boys to the bus stop in Ohio City. She was slightly startled to see me approaching with a rifle slung over my shoulder. She was more than startled to learn there was a lion in their shed. Hurrying the boys into the car, she sped off, shouting out the window that Curtis was in the house.

I quickly explained my predicament and asked whether he had a good flashlight. "I'll bring it down after I get dressed," he answered as I turned to hurry back to guard the only escape route again. A few minutes later, Curtis was ready to help me on my hunt. I stood facing the doorless shed, waiting for a beam of light to illuminate my target. "Batteries are dead!" he whispered in my ear. Not wanting to dawn my LED "candle" again, I inquired whether there was a light in the shed. There was, but it was a chain-pull switch near the back of the shed.

Now what? Curtis offered to lay out some extension cords and bring a halogen work lamp to the party. I nodded in agreement, not taking my eyes off the shed. Within a few minutes, I heard the unmistakable sound of ice being chopped. I turned, and sure enough, he was chopping on a mound of ice, under which was his extension cord. A few more chops were followed by a string of curse words.

"Did you chop the cord?" I asked.

"Yep!" he chirped.

Time for plan C or D. Plan C was to just back away from the shed, hide behind a tree, and see whether the cat would move. Plan C was for Curtis to hide behind the cement mixer nearby. Most typical mountain lion hunts involve using dogs and chasing the cats up trees; they don't usually include porches, sheds, and cement mixers. We shivered while we waited, keeping a close eye on the shed's opening and the light it allowed into the small wooden building. About five minutes were all it took, and I could hear rustling noises from inside.

The lion shot out from behind the tiny closet door, across the motorcycle and lawnmower, and up onto the stack of tires in the corner. We both had seen the lion, but there was no opportunity for a shot. It occurred to me then that I should inquire about collateral damage. "What's in the shed that you don't want shot?" I asked.

His reply was quick and dogmatic, "Nothing! I don't care if you hit something. I want this lion dead! I was in this shed last night at ten o'clock, and I have kids and a dog." I got the picture.

I could make out the outline of the cat and could tell how he faced on his perch. I checked again with Curtis. "I might blow a hole in your roof."

"I don't care. I'll fix it if you do!" he fired back.

I prepped for the shot with one more piece of advice for Curtis. "Okay, here goes. Stay behind that mixer in case it decides to bolt out the door. Let's give it somewhere to go other than over you."

He nodded at me, but his stare never left the doorway I was aiming at.

BOOM!

The recoil from the gun caused me to momentarily lose my visual on the cat. Then motion brought my gaze back to the stack of tires in the corner as the lion spun in a circle. I reloaded. By the time I had closed the bolt on the rifle again, the cat tore back across the mower and cycle and disappeared into the closet.

I knew I had hit my target but wasn't exactly sure where. So either there was a dead lion in the closet of this shed, or there was a wounded lion in that darkness. Hmm! I was getting tired of the "Now whats?" But that's exactly what I was thinking again. Now it didn't matter that we had no light, since the cat wasn't visible anyway.

We stood there for another few minutes, and I decided to creep into the shed and up to the wall of the closet. I didn't like being that far into the cat's territory, but I had to do something. I figured maybe I could hear his breathing, some indication of whether this cat was alive or dead.

This was now the second time in an hour I was putting myself within close quarters of a wild carnivore that could kill a five-hundred-pound elk. There was no turning back now; the cat deserved a quick and painless death, which Curtis and I wanted to help with. But our safety had to come first.

I moved extremely slow to avoid startling him if he was still alive. Every sense in me was fully alive right then, fully tuned in to my surroundings. Curtis's senses were as well. Once inside the shed and close to the closet wall, I stopped and listened. I could make out what sounded like shallow breathing. This cat was now within two feet of me, and the only thing separating us was a half-inch tongue and groove wall. I reached forward and with my fist knocked like I was at a stranger at the front door. I was hoping to hear nothing; the green light would allow me the courage to go further into this mess I had gotten myself in.

My knock was answered by a loud hissing, similar to an angry house cat. I'm not sure my feet touched the ground, but I was back behind my tree without remembering how I'd gotten there.

I glanced at Curtis. His eyes were huge, and I think mine were the size of golf balls. "He ain't dead!" I blurted out as our eyes met.

"I know!" he quipped back.

The adrenaline dump had been going on for over an hour now, and there wasn't an end in sight. I needed to finish off this cat for all of us, but how? Curtis had the solution. He grabbed a battery-powered skill saw from his shop and headed to the backside of the shed. I knew what he was going to do then. He planned to cut a small hole in the wall so I could safely shoot the lion. The challenge would still be darkness and distance, too dark and too close. But if I could see the cat, I might be able to shoot through the hole. Wounded animals are kind of like wounded people; they can hurt you. Normal behavior can no longer be expected. Fear and survival instinct change everything.

The hole was about head high and only five inches by ten inches, not big enough for the wounded lion to get through but big enough to maneuver the headlamp and rifle. It was pitch dark inside as I cautiously approached the wall and freshly cut window. I knew I still needed to see exactly where he was and his position. With the headlamp in my left hand and rifle in my right, I peered in.

Deja vu! Been there, done that, and I didn't want to go back. The light needed to be inside to be useful, exposing my left hand to a swipe if the cat wanted to. Two eyes reflected back at me, revealing his location. Then, as my eyes adjusted, I could see his body and where I needed to position the rifle. A three-feet shot should be easy, unless you can't see your sights and are worried about lion claws flying at you. Thankfully, my shot had been directly into vitals. I yanked the headlamp and rifle and, more importantly, my hands back out of the hole with catlike reflexes. A burst of commotion ensued that was unseen, followed by silence.

By now, Curtis and I were cold and in no hurry to go back into the shed. I suggested we warm up, get a hot cup of coffee, and ensure the lion was dead before we attempted to retrieve it. We sipped and warmed up, and I finally had a chance to tell Curtis the story that had led to me chasing a lion into his shed.

One more adrenaline rush still had to come. I needed to crawl over the lawnmower and debris, and hopefully find a dead lion. This time there was no hiss answering my knock on the wall of the small room. We were able to drag the beast out and into the light. We could finally see this cat. It was a three-year-old male, about 120 pounds. Hunters often take photos of their harvest, and we needed to as well. We each held a paw with the cat suspended between us with the shed in the background. Not exactly the cover photo for an issue of *Outdoor Life*.

*Curtis and I with the elusive lion in front of the shed.*

This adventure took turns I never saw coming. The adrenaline rush lasted way too long, and combined with the cold, left me exhausted. I coughed for nearly two months, knowing I had minor lung damage. I learned lessons that day I won't soon forget. For one, don't try sprinting through two feet of snow when it's negative 25 degrees. Get a better and brighter headlamp and improvise with skill saws when necessary.

Chapter 20

# Don't Run. Hide!

My dear friend Bill, who was about forty years older than I, always had some great encouragement for me. His family's cabin is next to our house in Ohio City. He and his sweet wife, Patti, came every summer and spent a few months. They were our part-time neighbors from Louisiana and more like family than friends. Bill often invited me over for "porch time." This consisted of slow rocking in a chair and drinking Community coffee, which he brewed more like a solid than a liquid. As we sipped the sludge, he often started the conversation the same way in his southern Louisiana drawl.

"Spencer, I wonder what the poor folk are doing today." This was his way of declaring his gratitude for the beautiful mountain view and the fact that we were able to enjoy it. After the conversation appetizer was finished, he moved on to the main course. Some nugget of wisdom he wanted to share with me or maybe a slightly off-color joke. We had some great talks, and I learned plenty while sitting on his porch in that rocker. He usually prefaced any advice he gave with, "Spencer, there are two kinds of fools: those who give advice and those who take it. I'll be one if you'll be the other." Then he shared some tidbit that was often applicable to someone in my season of life. These weren't universal truths but more common sense that had specific applications at times. One such piece of advice was "Don't hide. Run, boy!" The context for his advice has long been forgotten, but I do remember his comment and his demeanor. His accent seemed to help punctuate his authority. "Don't hide. Just run!" seemed like good advice for many dangerous situations…. until I was instructed to do the opposite.

"Don't run. Hide!" has been shared with me on several occasions. It really hadn't been given as theoretical advice like Bill's but more like a command in a couple of situations. The first time this was drilled into me was in Zimbabwe. A dear friend had invited me to join him on an African safari. Jon and I had shared some great adventures and developed a deep friendship over the decades, so when he called and announced that he was going on a safari and I was going with him, I didn't argue. Jon invited me to join him and capture what I could with the camera. This was a great gift to me, and I knew I was along more for companionship than for photographic expertise. In fact, he would need to give me a crash course with his 35 mm camera on the airplane if I was to capture anything at all. The only camera I had was similar to the disposable ones. With good equipment and instruction, I managed to press the button and blow through numerous rolls of film, with nearly half of the images in focus.

Since I had worked as a guide in Colorado and Alaska, I figured I was savvy and could adapt to the challenges a new continent offered. The main challenge was getting close to dangerous game.

This safari's focus was primarily elephants. Because we were in truly wild country, the elephants were quite wary of humans. This was not a zoo. So, there was a need to cover as much country as possible when searching for elephants and with some luck locate mature bulls. A mature bull elephant was based on the size of his tusks. The goal was to get close to a bull with long and heavy tusks, neither of which was broken. This is about where the similarities between searching for game in Colorado and Africa end. Of course, we had a guide in Zimbabwe. He set up the whole safari experience, hired nationals to pack our gear, cook and drive.

The others joining us during the ten-day safari were from local tribes. This group had experience packing gear and tracking game.

One of the locals was a government official. He wore a uniform and sported an AK-47 rifle with a ten-round clip slung over the shoulder. This semi-automatic rifle was for protection and to ensure that no poaching took place.

An African safari is an experience like nothing else on earth. The sights, sounds, and smells are all otherworldly. We camped along the Zambezi River in northern Zimbabwe. Hippos and crocodiles were plentiful and could be seen from our base camp. Each day we drove on rough roads deep into the bush, looking for tracks and searching huge areas with binoculars looking for game.

There is a good reason why some of the game in Africa is considered dangerous, and each one can pose a unique threat. On day one of the safari, our guide had us attend his safety course on elephants. The course was the "scared straight" version of life with the largest land animal in the world. Seeing docile elephants at the zoo eating bales of hay behind bars does more harm than good if getting close-up photos of one is on your bucket list. Wild elephants are smart and fast; they have an incredible sense of smell and know how to defend themselves. During our one-hour course, our professor and guide made the dangers involved very clear.

The illustrations and stories he told helped me erase every Dumbo circus image in my brain. He explained three primary ways adult elephants dispatch threats. Each of these methods started with the caveat "If an elephant sees you …" The first method involves grabbing the threat—in this case you—with the trunk and impaling you on a tusk. Numerous Africans have observed this method in various countries. The next is grabbing you and slamming you into a hard surface, including a rock or tree. The final one is holding you down on the ground with the trunk and squashing you under a front foot. All three methods are equally effective, rendering the same result.

So, the moral to these stories is this: if an elephant charges you, "Don't panic and run! Look for a place to hide!" Basic instinct tells us we can outrun these obese mammals. This instinct has been proven wrong on numerous occasions. However, elephants don't have superb eyesight—hence, the reason for the advice to never run. Movement makes someone easier to be spotted, so it's best to hide and be visible as little as possible. All this is applicable only in the event of a charge. If an elephant is startled and charges, try to find a place to hide but never lose your head and run. Our guide went over this instruction like your mother went over "Don't play in the street!" Okay, we've got it. We won't panic and run.

We spent days driving, hiking, and looking at elephants from a distance. It wasn't always easy to determine just how big a bull was, especially if it was a mile or more away. Because of obstructions and uncooperative elephants, there were times when we just needed to hike closer to the bull to judge his size. It was the dry season in Zimbabwe, and most of the smaller rivers and creeks were dry and sandy. This made these hikes much easier and faster. One of us had spotted a bull, which was well over a mile away and on the other side of a big, dry riverbed. We studied this elephant for at least fifteen minutes and still couldn't get a good view of his tusks.

Elephants spend much of their day feeding, especially in the dry season when some vegetation is harder to find. As a result, this bull was methodically reaching up with his trunk and pulling trees over until they uprooted. Then he was able to eat the leaves off the top, which had been unreachable moments before. He was fascinating to watch, and I was mesmerized by the steady approach in which it destroyed and then consumed a tree. The guide finally said we needed to get a closer look at this bull and would need to cross the valley, climb up the dry riverbank, and see just how big he was. So off we went on another thirty-minute detour. The guide, Jon, and I led the way, weaving through the brush and trees. If you have ever seen the *Lion King* movie, you know about the "tree of life." These huge trees with enormous trunks are called baobabs. Near the base, these monsters may be eight to ten feet in diameter.

The group of Zimbabwean trackers and packers weren't far behind us, but we never heard them. They were silent as we all stalked to the far side of the dry riverbed. The bank was steep and about twenty feet tall. Soft dirt and sand caved away under our feet as we slowly made our way up the bank so we could peek over. The bull hadn't moved far from his last tree toppling breakfast, and now stood motionless, facing away from us. Even without the desired broadside view, we could all tell immediately this wasn't a mature bull.

The sun was now high in the sky, and even though it was technically winter, it was getting hot. The guide explained it was nap time for this bull after a hearty meal, and he was sound asleep. Elephants can sleep standing up, and this one stood completely still with his giant butt a few feet from a baobab tree. The elephant and the baobab tree looked similar in size and color, both gray and about five to six feet wide. We all slowly made our way up to the top of the bank and now were only forty yards or so from the huge beast.

I clicked a few photos, but the only thought going through my head was *Don't panic and run; look for a place to hide. Don't run! Hide.* I rehearsed this mantra several more times, even mouthing the words. Since the elephant was asleep and facing away from us, and since I knew it couldn't charge us in reverse, I felt a little more at ease.

It then occurred to me that this was a great opportunity for a once-in-a-lifetime photo. I asked our guide if I could sneak up to the baobab tree, keeping it between the bull and me. I would leave the camera with Jon and have him take the pic. This was before digital photography and Photoshop, so this photo would be priceless. I'm not sure why he agreed, maybe he was hoping for a bigger tip if we got the photos we wanted. Whatever his reason, he agreed but only if he accompanied me. The two of us tip-toed until we reached the baobab tree. It is a strange feeling to be that close to any wild animal but especially one with that size and potential.

Jon captured the photo perfectly. We crept back, me with a huge smile that said it all. I got a big thumbs-up from the group of guys standing about twenty-five feet behind Jon. He whispered that it looked awesome, and now it was his turn. He also wanted to be in this stunning scene that close to an elephant. So, up to the tree Jon and the guide went. I was clicking away as they both posed, smiling and pointing through the tree at the bull on the other side. Jon motioned that he wanted one more photo. *Click! Click!*

There wasn't a breath of wind that day, not even a small rustle in the leaves. Nothing! It was dead silent, except for that Nikon click, which now seemed obnoxiously loud. The next sound I heard came simultaneously with the spin movement. It was like being very close to a lightning strike, where the flash of light and the deafening crack are inseparable.

The bull whirled around in one fluid motion. As he did, his trunk recoiled close to his face and his ears flared out from the sides of his head. The trumpeting sound reverberated and seemed to come from all around, not just from his trunk. Details are still poignant. His eyes were wide with fear or terror or anger.

In a split second and in two strides, he was up to his top-end speed, which was now in my direction. *Don't run. Hide! Don't run. Hide!* I remember thinking, *Hide behind what?* Jon and the guide were completely hidden from the bull by the massive baobab. I stood in the wide open without a tree, rock, or even a bush to duck behind. The only thing between me and ten thousand pounds of fury was a 35 mm camera. *Don't run. Hide!* The advice was now fleeting as I too whirled and spun 180 degrees.

I'm not sure what I expected to see, since my field of view abruptly changed, but what I saw wasn't even on my radar. Ten Zimbabweans had not only disregarded the guide's tutorial but also had beaten me to the punch.

There was only a blur as all ten of them ran wildly without a thought of hiding. They simultaneously scrambled off the steep bank and back down to the riverbed. Not only was I probably the slowest in the bunch, but I didn't have the head start each of them had. There is no possible way this poor-sighted elephant could miss the movement of ten Africans running wildly, followed by a slower white guy. Adrenaline kicked in and provided the boost I needed. I wasn't going to catch the others who had head starts, but I was making some anti-gravity-type strides to the embankment, which I long-jumped off.

I hit the riverbed in three steps with the Nikon flying around my neck and bouncing off my chest. I felt no pain, even though it was bruising my sternum. I covered another fifty yards in about five seconds and caught up to the group of cheaters. They now faced me and seemed to be cheering me on.

Suddenly, I saw the barrel of the AK-47 rise. I turned back just in time to see the massive bull towering over us at the edge of the steep bank, ears still flared out and eyes still wide. I'm not sure whether the bull decided it was too steep for him to come down, or he figured he had made his point. Either way, I heard, "Crack! Crack! Crack!" Three rounds from the AK were sent just over the bull's head. With the same speed, the bull whirled away from us and disappeared out of view.

The packers, trackers, and I began to high-five each other and hug like we had just won some championship game. Jon and the guide appeared at the top of the bank, shaking their heads in disbelief. Every time I look at the photos of Jon or me, the elephant, and a baobab tree, I still get a tiny adrenaline rush. At that moment, I was just thankful to be alive and not skewered by a tusk. But later, as the rush subsided, I couldn't help but think of the advice to "not run and hide" and wondered whether they encouraged only the non-Africans to do this. I think the advice given to the Zimbabweans was, "Run! Don't worry if you aren't faster than the elephant. Don't even worry if you are not the fastest one in the group. You just don't want to be slowest one in the group. So, Run! If you are slow, make sure you have a good head start!"

Zimbabwe wasn't the only African country where this advice was shared with me. South Sudan was another place where I was compelled to "hide and not run." In 2011, South Sudan became the world's newest country. In 2010, Annette and I had made our first trip into the town of Juba and then north to a tiny village, where a Bible college had been set up. The people in this part of Africa are well acquainted with war, and it has gotten only worse in recent years. During the early 2000s, the Muslim government in Khartoum attempted to wipe out Christians in the southern part of Sudan. As a result, many pastors had been killed, and churches burned.

The Bible college in Heiban was set up to train and equip new pastors and rebuild churches. We had the privilege of leading a retreat for the international staff that gathered from across the southern part of what was then Sudan. The following year, under the guidance of the United Nations, a boundary had been established, and the infant state of South Sudan was born.

All this did little to stop the ongoing war and the government in Khartoum from dropping bombs. In an effort to rid the north of everyone but Muslims, bombing raids created numerous refugees streaming south. Camps had been set up to receive and help refugees in several places near the new border. On this 2011 trip, I would facilitate a retreat for relief workers at a camp in the northeastern part of South Sudan. I flew on commercial flights as far as Juba, the new capital, and on smaller missionary planes from there to the refugee camp. From the air, these camps are quite visible with blue-and-white United Nations tents or tarps dotting the landscape. This camp was strategically located about ten miles south of the border and hopefully far enough to escape the continued bombing.

Sheep and goats had to be shooed off the new runway before we could land. A wall of heat was my first greeting as I stepped off the small aircraft. It was suffocating. One hundred and twenty-two degrees and 90 percent humidity felt like I had closed a sauna door behind me. Once again, I figured if these people could live here for months on end, taking care of refugees, the very least I could do was come and do a retreat for them. Usually, when I did a retreat like this, we took them away from their work areas for four or five days. But with a constant and daily stream of new arrivals into the camp, there was time only for devotions each morning and then being available to meet individually at other times. The needs were growing with each new refugee family for water, food, shelter, and medical treatment.

The first evening I was shown a canvas army tent, which would be my quarters for the next week. The heat inside the dark tent was stifling. Outside, the sun was nearly down, and it still felt like an oven. I drank water, bottle after bottle, and never seemed to urinate. I threw my bag on the cot and stepped back outside. Next to my tent and every tent was a hole in the ground. They were round and about three feet wide and three feet deep. I realized then that I had my own designated foxhole. Don't run. Hide! I flashed back to the same advice years earlier given to me in Zimbabwe. *If the bombing gets close, don't run; hide in your foxhole.* Somehow, I knew I wouldn't need to try to remember the advice this time. In a matter of seconds, it was now etched on my brain. Thankfully, the bombing never got close the week I was there. I didn't once have to use the foxhole.

Early the next morning over breakfast, I was introduced to the staff and led them in a rushed devotional. It was clear that everyone had a pressing task after breakfast and would soon be on his or her way, delivering water or food or putting up more tarps for makeshift tents. Others would head over to a metal building, which had been thrown up over the last few weeks to be the hospital. Because of the bombing from the north and the trek many had to make through minefields, there were many who needed medical treatment. One of the medical personnel was a doctor, who had spent much of his professional life in parts of Africa like this. He was operating on people in this tin shed. I remember thinking that this made a *M\*A\*S\*H* unit look like the Mayo Clinic.

As I finished the devotional and prayed, most were finishing breakfast. Just before the doctor took off, he asked me whether I knew anything about hydraulics. I thought it was a strange question but thought for a second and responded, "My dad has a tractor and backhoe. Does that count?"

He said, "Yep, it does. Come on." I followed him as he walked briskly toward the make-shift hospital, and the line of people waiting to see him. The building was divided into two rooms by a plywood wall, which didn't go all the way to the metal ceiling. Inside one room was a plywood table and one chair with a hinged piece of plywood for a door. There were a few metal cabinets stocked with basic medical supplies, mostly first-aid supplies. On the other side of the wall was my destination, the "operating room." There was a piece of plexiglass in the ceiling used as a skylight and also a real OR lamp, not yet connected to electricity. Under them was my project for the day.

A brand-new, half-assembled operating table sat in the middle of the concrete floor. The new white paint on the floor was contaminated with hydraulic oil. Numerous stainless-steel table parts were strewn about along with a handful of tools and oil-soaked pages of the assembly instructions. *It's a good thing I like puzzles*, I thought. It took the better part of the day, but eventually, the puzzle was built. In the end, an OR table complete with foot pumps and all the tilting, raising, and lowering capabilities was in working order. I felt accomplished and thankful my dad had some tractors. Now anytime I hear something remotely similar to the advice of "Don't run. Hide," my mind goes straight to another continent. Africa is a place to hide and not run.

Advice is a tricky thing. I often think back to my friend Bill's. I think he knew his nugget of wisdom could be reversed given the right set of circumstances. I have found there are occasions when both can be true at the same time. When it comes to the adventure of walking with God, both are definitely true. There are times when we may be tempted to run from God, but what we really need is to hide in him. God will be a refuge in times of trouble and tucking myself away in his presence is the safest place to be. On the other side of this coin, however, is another truth. There are times when the temptation is great to hide, when I actually need to run to him. Either way, the common denominator is God, whether we need to run or hide.

CHAPTER 21

# Wanna Go to Iraq?

We called them doors. That's really what they seemed to us. Doors were opportunities to step into other adventures while being led by God. We didn't say yes to every one of them, but if the door was still open after praying, we often went through it. Of course, depending on the destination, some people thought we were nuts. We have been privileged to work with several organizations and help facilitate retreats for missionaries and relief workers all over the world.

Ears perk up when you tell someone you are going to Sudan, Myanmar, or Cambodia. Often the retreats we do are for relief workers, who are ministering to refugees or people affected by natural disasters. We continue to say that if there are people who are willing to live and serve in these places, then we certainly should be willing to go and spend a week with them and hopefully refresh them. Still, we have become accustomed to someone's look of shock when they hear some of the destinations, especially when we took our daughter Kate with us to help with childcare.

"What? You are taking your wife and daughter to Iraq?" This seemed to be the most common response.

One of my first directors with CRU and his wife had raised their children in many countries, where they were working as missionaries, places that didn't offer travel guides or any trip advisor. When he was questioned about the safety concerns of taking his children to these places, he usually answered that the safest place for his family was in the center of God's will. I have never forgotten his response and have used it a few times myself. If people weren't shocked about where we were working or traveling, then it usually meant one of two things: either they wanted to join us and fully understood what we were doing, or they were convinced it was a cover for something else. A dear friend's dad was, and maybe still is, convinced that I work for the CIA. Of course, we also realized that not everyone looked through the same lens we did or had the same thirst for adventure. I was just thankful that Annette and I both had it and were able to pass it on to Andrew and Kate too.

I have made seven trips to Iraq since 2012; Annette has been with me on six of those, and Kate has joined us twice. We have dear friends who live there and are raising their family there, so again, the least we can do is take a week once a year and go to be with them. Each time has been a privilege and an adventure. We try to avoid any of the hotspots where there is fighting and are usually in the northern part of Iraq, known as Kurdistan. Most Kurdish people are very welcoming toward Americans and friendly. They are also thankful for the US-led invasion into Iraq, which ousted Saddam Hussein, since he was committing genocide against the Kurds. So, as we rolled through checkpoints on highways and flashed a US passport, the armed soldiers usually smiled wide and gave us a big thumbs-up. I have felt quite safe in northern Iraq.

The one trip when Annette wasn't with me was when my friend in-country asked whether I would come over and help distribute aid to children in refugee camps. Of course, I smiled and gave him a thumbs-up. At the time, there were over half a million displaced people in Northern Iraq. I recruited a friend to join me to help with the distribution. We had our itinerary, and we all packed. I would fly out of Gunnison, Colorado, and into Denver, where we would meet. Then we were on to Frankfurt, Istanbul, and finally Erbil, Iraq. Once again, it's the unexpected that adds so much to any adventure. It's the little things we didn't see coming or weren't ready for that made the ordinary trip to Iraq into tales that might be told for years to come.

My friend and travel partner texted me two days before we were to leave. That evening he was feeling sick and just wanted me to know so I could pray. He would hopefully feel better the next day and let me know. Several months of planning and purchases were in jeopardy. The unexpected had hit. He was sick and had vertigo, not a good combo if you are scheduled to get on multiple flights. I was committed to going with or without him, but I wasn't looking forward to going alone. I have traveled solo many times internationally, and it's always more fun with a friend. In less than twenty-four hours, I would be on my first flight.

The wheels began to turn. Whom could I recruit with such short notice? Who would say yes to Iraq? I had a flashback to twenty-some years before, when I was trying to recruit someone to move to Mongolia for a year. At least this time it was only a ten-day trip.

My first option was my brother. "Hey Matt, wanna go to Iraq? Tomorrow? For ten days?"

Since he was the hatchery chief for the State of Colorado with Parks and Wildlife, it was a long shot, but I tried. I moved to another friend and board member with Big Horn Ministries. His passport had expired. This isn't one of those situations where the third or fourth person you ask feels slighted or overlooked because he wasn't first on the list. In fact, it wasn't even a situation where someone thanked you for thinking of him.

It was now midday, and the clock was ticking when Annette suggested Doug, a dear friend for decades and someone who had lived in Chile for nine years with his family, working as missionaries. Yes, Doug was savvy and currently working on his doctorate in theology, which meant he probably needed a break from writing his dissertation.

I started with a text. "Bro, wanna go to Iraq … tomorrow?" I followed this up with a phone call and a longer explanation. Surprisingly, he stayed on the phone, asked a few questions about tickets and the layover in Istanbul on the way home, and most importantly didn't hang up. I could tell he was intrigued, and I had appealed to his sense of adventure. He said he would pray and talk to his wife as soon as she got home. This gave me a little hope but also caused a little more anxiety since she wouldn't be home for a few hours. By then, the recruitment window would be just about closed.

Here is my new definition of a good friend. It is a person you can call, with about twenty hours' notice, and invite to join you to visit refugee camps in Iraq, and the person says yes. That's a good friend.

It's a good thing that Doug looks nothing like a guy who wants to join ISIS or become a Jihadist, because when someone buys a ticket and travels to Iraq the same day, it raises eyebrows. The unexpected parts of this adventure served to deepen a friendship and make for a lot more than just traveling together. The time spent distributing aid inside the refugee camps was also significant and has had a lasting impact on me. The screen saver on my computer is a photo taken from that trip. I am with a Syrian father and his two young daughters, who each have an armload of boxes and winter coats. I am reminded of that trip nearly every day.

Friendship is also deepened through understanding. This trip was a time to be better understood by a friend. We all need to be understood at some level by others. Understanding doesn't come easily; it takes time with people who care about you and you feel safe around. After this trip with Doug, I felt understood in ways that cemented our friendship. After eight days in Iraq, we departed for Istanbul, where we planned to spend a couple of days seeing the sights. It is a beautiful city, rich in history, and there is much to see. We planned to visit some of the major attractions, enjoy a restaurant or two, and maybe even relax at a Turkish bath. We had heard the Turkish version of a sauna was a must, if you wanted to relax and go a bit further into Turkish culture. After a few tourist stops and restaurants, we found a public bathhouse near the hotel.

We had both been to a number of public saunas in Russia or other countries and actually considered ourselves to be veterans at cross-cultural public bathing. The basics are likely the same: steam, relax, sit in quietness, and stay hydrated. We got this! We confidently headed for the men's bathhouse. To our surprise, there were several menu options and varying prices. If you have ever spent longer than an hour in a sauna, you spent too long and came out like a prune. One hour should be sufficient, but there was also the option of a massage. I paid the extra for the massage, and we then received our towels, robes, and slippers. Off to the locker room, we went to change.

*Doug and I entering the Turkish bath - the last moment of relaxation.*

There is one major difference between public saunas or spas in the United States and everywhere else in the world. The rest of the world goes naked, assuming it isn't coed. Most Americans, including me, are uncomfortable with this and end up wearing the towel or robe for as long as possible. When it's too awkward to wear any longer, we then incorporate the fig leaf pose by crossing our hands in front of the groin area. But eventually, buck wins, and you must be naked.

We dropped everything in the entry room and proceeded with the fig leaf pose into a huge room made entirely of marble. Around the perimeter were about a half dozen showers, not stalls, but there were showerheads near the floor with a five-gallon bucket next to each. In the center of the room was a circular bench also made of marble. There were a handful of men relaxing in different locations around the room. There were also several big doorways exiting the main room. These led to the steam rooms, where more men were sitting and relaxing, eyes closed and breathing deeply.

We entered one steamy doorway and disappeared from each other for a few minutes. We found seats and began to relax more as we dropped the fig leaves and our guards in the mist. Ten minutes was my max, and I had to find some fresh, cooler air. After a few ins and outs, a toweled Turk stood in the doorway, motioning me to follow him. I made out the word *massage*, what I had been anticipating. He was a bulky block of a man with more hair than I have ever seen on a human being. His mustache looked like a ferret, which had fallen asleep on his upper lip.

I suddenly felt bald, extremely naked, and I reconsidered my massage. I followed him like a kid who follows the school principal. Once in the central room, he walked to the marble bench and spread a towel. Then, without a smile or a hint of customer service, he motioned for me to lie facedown on the towel. I immediately did as I was told, happy to be facedown and not eye to eye with my manly masseuse.

That day, there was a very fine line between a relaxing massage and abuse. That line is the boundary of personal perspective. I'm sure from the Turk's perspective, it was a deep, deep tissue massage. I, on the other hand, felt like a boy in a man's world I wanted to escape but didn't know how. And this was before I was instructed to turn over!

Part two of this torture lasted only about ten minutes but felt like hours. Finally, without warning or any international sign language, he grabbed my arm and pulled me upright. With a grip a cop uses on a criminal, he marched me over to one of the showerheads. He motioned me to sit on a step. Cowering in fear, I did. Exposed was the main thought rolling around in my sweat-soaked head. *I am totally exposed!* I was so tense that relaxation would likely not come for hours. I wasn't sure what to do and was also totally oblivious to what the Turk was doing.

*Splash!* Five gallons of warm water hit my head, jolting me momentarily from the fixation on being exposed. I glanced over as he set down the bucket and grabbed a brick-sized bar of soap. *No! No! No!* Too late. I was back to fixating on being a bald boy in this man's world and getting lathered up by the hairy hulk. I may have let out a whimper at some point. I don't remember. I have tried to block as much of this experience out of my memory as I can. Thankfully, the soap suds stopped at my waistline. Before I could wipe the soap from my eyes, another forceful five gallons of water hit me.

Unfortunately or fortunately, depending on that thin line of perspective, there were no witnesses to this crime. Through the fog, as if by magic, suddenly Doug was standing next to us. I was glad to see my friend but also curious as to what he had seen. The hulk seized the moment, and the next thing I knew, my naked friend was sitting next to me. He started to ask me about the massage, but I knew I was suffering from mild PTSD and wouldn't be able to formulate any words.

Then I heard the bucket filling again and braced myself for round two. With a quick motion, the Turk drenched Doug as I stared, which gave me a vision of what I must've looked like minutes before. Shocked and an open mouth gasping for air. Even if I had had the time and words to tell Doug what was coming, I'm not sure I would have. Sometimes the anticipation makes the event worse. Best to let Doug just endure it as it came. He also was confused by the huge bar of soap being quickly rotated in the hulk's hands, creating a lather. In less than a second, Doug was sudsy to the waist. About then I was able to choke out two somewhat legible pieces of advice: "Bro, don't stand up" and "Just let it happen."

Doug's gasp and sudden crouch were the signs I was waiting for. The second five gallons of water ran off him, and we both turned to see the Turk drop the bucket and mosey off. It was over! We had made it! There was no need for words or an explanation right then. We sat in silence for a few minutes, staring at the floor. I gave him a few minutes for the shock to wear off and then broke the solemnity of the moment. "Bro, I have never felt more understood in my life." He just nodded. Our friendship was deepened that day through shared trauma and understanding, all because Doug had stepped into the adventure with me and accepted the invitation to go to Iraq.

CHAPTER 22

# The Adventurer

In general, Americans have a fascination with automobiles and the possibilities they bring. Car manufacturers even try to define us by the products we buy. In recent years, many makers of SUVs have given names to their four-wheel-drive vehicles, which perpetuate this narrative. Names like Explorer, Expedition, and Wrangler help move us away from our mundane lives and call us into the unknown. If these names don't move you to purchase a car, then places like Durango, Rubicon, and Outback might persuade you. If those still don't motivate you, then maybe Xterra or Crosstrek will.

The marketing simply taps our longing for adventure, and we're led to believe this adventure must be impossible unless we own one of these autos.

My parents have purchased only one new car in the last fifty years. In 1975, they decided our family needed an all-purpose, four-wheel-drive vehicle. We needed space for five people, which most cars had during that decade. This vehicle would also need to double as a work truck for my dad, pull a horse trailer, and eventually become a snowplow.

The 1975 Dodge Power Wagon pickup was Dad's choice. This pickup truck still sits at the Big Horn Guest Ranch in Ohio City, where we live. It has a snowplow attached to the front and about twelve hundred pounds of steel in the bed to help with traction. It also has chains on all four tires, which have not been removed for about a decade. The odometer, which still works, shows just over one hundred four thousand miles. But odometers don't work when cars go in reverse (ask Ferris Bueller), and when one is plowing snow, reverse is used nearly as much as forward. So the truck probably has closer to two hundred thousand miles on it. I have plowed a lot of snow with this old Dodge, especially during the winter of 2019.

I was plowing one day early in the winter, and after a few hours of mind-numbing back-and-forth, I lost what little focus I needed. It was then that I sideswiped a fence post, adding one more dent to the driver's side.

I didn't know it then, but a decorative metal emblem fell off and was buried in a four-foot snowbank. The shiny metal was slowly revealed by the receding snow in May. At first, I couldn't remember which of the relics it had originated from. Then I remembered the scraping sound back in January in that same vicinity. "Adventurer" was the Chrysler Corporation's early attempt to market this gas-guzzling power wagon. But despite the perfect name given to this truck, I don't remember having many adventures in it.

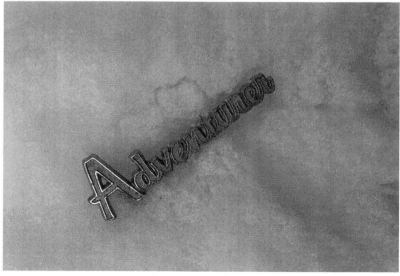

*Snow melt revealed what should have never been on that work truck.*

In fact, this truck is synonymous with work in our family. The only thing I do in this truck is work. The only thing anyone does in this truck is work. How fitting that the metal emblem and once-valuable marketing effort, which had recently fallen off and was lost in the snow, while it was working. Cosmetic care for this car had ended about forty years ago, but I did save the metallic tag written in cursive. The marketing may have worked on my dad and coaxed him into buying the only brand-new vehicle of his life, but it was mislabeled, proven by the fact that the old 1975 Dodge hasn't seen an adventure in over four decades.

What's my point? We can label something as an adventure, but that doesn't make it so. One definition for adventure *is a risky undertaking; a remarkable and exciting experience.* The only thing risky that happened in that old pickup over the last decade was when this driver whacked a fence post while backing up in a blizzard. The only remarkable thing about that pickup is how often you need to fill it up with fuel. Yep, labels can be misleading. Similarly, we don't need to use the word *adventure* to experience it.

The word *adventure* never appears in the Bible. Not once. And yet the Bible is packed with adventure. From Genesis to Revelation, all through scripture, adventure is never far away. God's incredible plan for mankind, which unfolds in the Bible, is an adventure in love. Another word that never appears in scripture is the word *Bible*. I am tempted to take the metal "Adventurer" emblem and glue it on the front cover of my Bible, the New International Version, open and read for adventure.

Love is the ultimate adventure. The risk, of course, isn't knowing whether it will be returned. The remarkable part of love is how it changes us. And the exciting aspect of love is when we give it away. Thus, love is the ultimate, containing all the requirements for adventure. The greatest adventure the human spirit can ever experience is the love of God. In fact, it is what each of us was made for: to know and experience the unconditional love of God. This love initiates and seeks, it changes us and pulls us close, and it's impossible not to share.

Over the last sixteen years, Annette and I have had the privilege of working for a small nonprofit ministry called Big Horn Ministries (BHM). The name is derived from my family's guest ranch. The entire purpose of BHM is to "help people take their next step." And the next step is prayerfully one that leads a person closer into relationship with God. So, we help with direction, maybe offer a hand to get back on his or her feet, maybe a shoulder to lean on or a hand to take the first step with. Like a father giving away a bride, our desire is to help place someone's hand in God's. We want to see people step into their own adventure, being led by the Lord.

## Conclusion – The Adventure Continues

Thanks for tagging along with me on these adventures. I'm sure you must have many of your own. And since I am only in my mid-fifties, I hope I have a few more chapters as well. I'm not sure where God will take me next, but I don't need to. There was a time in my life when I desperately wanted to know what was next. I still want to know, but not with the same desperation. Pieces of the planner and the controller in me have died. More still need to die. The wonder of letting the One who knows and sees all lead us into the future is part of the great adventure. At times, He will show you some of what is next or just enough to take another step. Having ears to hear His voice are important tools for walking into the future. The seeds of vision are sewn in fertile, rested fields. I pray that both you and I have such an intimate walk with God that it allows us to be fertile soil for his seeds of vision to be planted in. But, just like the farmer who occasionally rests a field, we too must be rested in order to hear. Busyness will choke out the voice of God and with it clutter the path of adventure He wants to lead us on.

One of my favorite verses is Acts 20:24. Paul writes to the elders in the church at Ephesus, "However I consider my life worth nothing to me, if only I may finish the race (adventure) and complete the task the Lord Jesus has given me—the task of testifying to the gospel of God's grace." May you be steadfast to finish strong His adventure for you!

I'd love to hear about your adventure. If you are inclined, tell me about it. spencer.nicholl@bighornministries.org